SUPER COSMOS

Through Struggles to the Stars

By

Jack Sarfatti

First published by AuthorHouse 05/02/06

ISBN: 1-4184-7663-3 (e)
ISBN: 1-4184-7662-5 (sc)

Library of Congress Control Number: 2004095148

Printed in the United States of America
Bloomington, Indiana

This book is printed on acid-free paper.

SUPER COSMOS

The Theory of Everything for Everyone
(Second Revised Edition)
By Theoretical Physicist

Jack Sarfatti

(Ph.D., M.S. in physics from the University of California, Cornell BA in physics)

Do we need God as the Intelligent Designer? Is there any truth to the sensational stories by the late Colonel Phillip J. Corso and others in the USG Intelligence Community that we have been visited and manipulated by alien intelligences in flying saucer time machines *from the future* with telepathic and psychokinetic powers of mind control? Are alien "Skinwalkers" invading Earth coming through wormholes connected to a parallel universe next door at a Utah ranch owned by a "Howard Hughes" Las Vegas hotel/aerospace tycoon? Or, is that only a silly irrational conspiracy theory by self-deluded New Age "kooks" and "paranoids" using what Richard Feynman called "cargo cult science"[i] magical thinking from a bad science fiction comic book? You, dear reader, will have to decide that for yourself. This book explains the physical principles of consciousness, subspace warp drive and wormhole star gate time travel in the light of the new precision cosmology of anti-gravitational repulsive dark energy. Can the fabric of space-time be manipulated directly by Q-Consciousness like in Star Trek and as reported by ex-US Air Force intelligence agents who you find out about in this book? Is that also really only a fantasy and disinformation? Newton realized that the fall of the apple was the same as the motion of the Earth around the Sun. That was his alchemical insight "As Above So Below." I have shown that zero point exotic vacuum dark matter forms the invisible galactic halo that holds the stars together in the same way that the electron's distributed repulsive electric charge is held together sucked in by a counter-intuitive strong gravity attraction created by positive quantum pressure. This prevents all electrically charged matter from exploding and destroying the entire universe. Upset that delicate balance and we are in deep trouble like Kurt Vonnegut's "Ice 9". We are like Mickey Mouse as "The Sorcerer's Apprentice" in Walt Disney's "Fantasia". Control of dark energy is the "metric engineering" of weightless warp drive and wormholes for super-fast travel in space and even backward through time. It's "As Above So Below" all over again. Real original physics on the deep structure of the many worlds of the self-creating megaverse of pocket universes on the "cosmic landscape" not found in any other book is weaved together with great literature and my true adventure story. Quantum reality is not only weirder than we have imagined, but may be weirder than we *can* imagine. We are such dreams that stuff is made from. "The Question is: What is The Question?"[ii]

SPACE-TIME AND BEYOND – THE SERIES

Episode III: Through Struggles to the Stars - Ad Astra

Table of Contents

Prologue

Dear Reader in this book you find the basic battle-tested laws of physics that will allow us to make Star Trek real and travel like Gods to the stars and beyond even backwards in time. The good news about the parallel universes theory, now favored by most of the top theoretical physicists at the best universities, is that everything you ever wanted you actually get in some universe. Indeed, you get it many times over and over. That elusive beautiful girl you once fleetingly saw that you wanted, you got her. The money you wanted, you got it. The bad news is that all the most terrible things you imagined also happen. This is the ultimate democracy. Everything that can happen does happen. There is a Catch 22 in that most physicists say that you cannot experience the parallel universes next door because of signal locality. However, in this book I argue that signal locality is violated in all living matter. It is only a property of dead matter. Living matter is a qualitatively different emergent phenomenon beyond the reductionism of signal locality. Shamans and "Skinwalkers"[iii] travel to other worlds routinely, or so they say.

"The Mother of All Physics Problems"

The vacuum energies of fermions and bosons do not cancel, and the bottom line is that our best theory of elementary particles predicts vacuum energy whose gravitational effects would be vastly too large. We don't know what to make of it. ... Finding the reason has been regarded as the biggest, most difficult problem of modern physics. No other phenomenon has puzzled physicists for as long as this one. Every attempt, be it quantum field theory or in String Theory, has failed. It truly is the mother of all physics problems. " p. 77-78, Lenny Susskind, "Cosmic Landscape."

"Creation without the Creator"?

"Lest I be accused of taking a cheap shot at string theory, or advocating a deistic view of the universe, consider the following creation story which, after John A. Wheeler, I shall call "Creation without the Creator". Many extrapolations of continued exponential growth in computing power envision a technological singularity in which super-intelligent computers designing their own successors rapidly approach the ultimate physical limits on computation. Such computers would be sufficiently powerful to run highly faithful simulations of complex worlds, including intelligent beings living within them which need not be aware they were inhabiting a simulation, but thought they were living at the "top level", who eventually passed through their own technological singularity, created their own simulated universes, populated them with intelligent beings who, in turn...world without end. Of course, each level of simulation imposes a speed penalty (though, perhaps not much in the case of quantum computation), but it's not apparent to the inhabitants of the simulation since their own perceived time scale is in units of the "clock rate" of the simulation.

If an intelligent civilisation develops to the point where it can build these simulated universes, will it do so? Of course it will—just look at the fascination crude video game simulations have for people today. Now imagine a simulation as rich as reality and unpredictable as tomorrow, actually creating an inhabited universe—who could resist? As unlimited computing power becomes commonplace, kids will create innovative universes and evolve them for billions of simulated years for science fair projects. Call the mean number of simulated universes created by intelligent civilisations in a given universe (whether top-level or itself simulated) the branching factor. If this is greater than one, and there is a single top-level non-simulated universe, then it will be outnumbered by simulated universes, which grow exponentially in numbers with the depth of the simulation. Hence, by the Copernican principle, or principle of mediocrity, we should expect to find ourselves in a simulated universe, since they vastly outnumber the single top-level one, which would be an exceptional place in the ensemble of real and simulated universes. Now here's the point: if, as we should expect from this argument, we do live in a simulated universe, then our universe is the product of intelligent design and Theory 1 is an absolutely correct description of its origin." John Walker, The complete article is at http://www.fourmilab.ch/fourmilog/archives/2006-03/000664.html

Indeed John speculates that it may be "Simulations all the way" somewhat like Gott's "self-creating megaverse" discussed below. That is, the programmer is part of his own simulation in a Godel self-referential bootstrap reminiscent of what Doug Hofstadter called a "strange loop" in "Godel, Escher, Bach." Frank Tipler's "The Physics of Immortality" is relevant to this issue. As this book goes to press we cite John Walker:

George Ellis on Messages from the Megaverse

"In my recent review of Leonard Susskind's The Cosmic Landscape, I stated that observational evidence for the existence of the multiple universes supposed to make up the megaverse was "barring some conceptual breakthrough equivalent to looking inside a black hole, forever hidden from science by an impenetrable horizon". This is an allusion to chapter 12, "The Black Hole War", which suggests that Stephen Hawking's throwing in the towel and conceding that black holes do not cause loss of quantum state information means that it may be possible (at least in theory) to obtain information about other parallel universes encoded in radiation emanating from the cosmic horizon which delimits the observable universe. George Ellis, who, along with Stephen Hawking, literally wrote the book about The Large Scale Structure of Space-Time has now posted a brief comment on ArXiv, "On Horizons and the Cosmic Landscape", which says that this is all nonsense, based on Susskind's confusing event horizons such as those which surround black holes with particle horizons, which are observer- and time-dependent, and simply specify the volume of space-time from which light has had sufficient time to reach an observer; your own horizon, which is slightly different from everyone else's, expands in every direction by one light-year from one birthday to the next. Ellis also observes, as did I, that the visual horizon due to the cosmic background radiation further reduces the volume accessible to observation and concludes, "The ESA-NASA Planck Surveyor data will not have coded into it the nature of multiverse regions enormously more distant from us than a Hubble radius. An application for a grant to decipher information about far distant regions of a megaverse that may be hidden in the data will not succeed." http://www.fourmilab.ch/fourmilog/archives/2006-03/000665.html

Of course, we do not have Susskind's response to this as this book goes to press. The point is that science is a process of discovery and error correction – the never-ending story. Furthermore, there is also the issue of "signal nonlocality" in violation of micro-quantum theory's "no-cloning theorem", though possibly not macro-quantum theory. This would allow us to receive and also send signals beyond all sorts of horizons.

The Rogue Scholars of The Invisible College

My thanks to all the people whose writings I cite below for their valuable contributions to this chronicle of the difficult birth of metric engineering warp drive and star gate travel to the stars and beyond. I also thank Angela Nissim and Kelvin. I also thank everyone whose names you see throughout this book for contributing to my understanding of the physics of "*Unconventional Flying Objects*" in the sense of Paul Hill's classic physics text on the subject. The physics in the book is still under construction emerging as in my two earlier books from 2002. It will get better with each successive book, but it is original and addresses the important unsolved problems of the day. Though I stumble on a few keys, there is Soul in my music. Note, I use the terms "subspace" and "substratum" interchangeably below in the sense of the Einstein-Cartan tetrad gravitational field. This book is in the genre of fantastic realism and science-faction like the classic "Morning of The Magicians" by Louis Pauwels and Jacques Bergier. The science is real. The alleged facts are real to the best of my knowledge. My interpretation is not conventional, but is strongly colored by my direct contacts with the high strangeness of UFOs and the paranormal. These contacts are described in my autobiography "Destiny Matrix." Like The Pirates of Penzance I come from a band of "Rogue Scholars" in what has been called the "Invisible College." I should mention Dan Smith the "phenomenologist" of The Aviary and The Aquarium that are legendary associations of current and former Spooks interested in weird phenomena beyond the fringe just in case it could result in technological surprise. This is included in the area of interest of MASINT[iv] in the new post-911 National Intelligence Directorate. Dan, allegedly with former USAF Intelligence agent Richard Doty and others like Christopher "Kit" Green, has been attempting to get official confirmation or denial of what Dan calls the "core story" in which alleged "EBE's" ("Extra Biological Entities") have made agreements with several intelligence agencies of several nations. Dan had a transformative experience (similar to Joe Firmage's) that cut short his promising physics career at Princeton, Stanford, University of Maryland etc. shortly after he read my 1975 book "Space-Time and Beyond" in 1977 co-authored with Fred Alan Wolf (star of the movie "What The Bleep") and artist Bob Toben. Hal Puthoff and Kit Green, along with Jacques Vallee and Col. John Alexander (who knew Colonel Phillip Corso), were also on the Science Advisory Board of Robert Bigelow's NIDS on whose ranch the Skinwalker was hunted.

The organizing idea of this book

Ed Witten of the Institute of Advanced Studies in Princeton laments the fact that Brian Greene's "Elegant Universe" of string M-Theory lacks an "organizing idea" and that the discovery of universally repulsive anti-gravitational "dark energy" is the greatest crisis in his brilliant career as the New Einstein. Nobel Physics Laureate Shelly Glashow harshly condemns string theory as un-testable pseudoscience making the thousands of theorists in our university physics departments into the Laputan academics lampooned in Jonathan Swift's "Gulliver's Travels."[v] The recent discovery of the "fireball" may auger a coming victory for string theory as does Lenny Susskind's new theory of the "cosmic landscape" populated by actual pocket universes continually created in eternal chaotic inflation. This "megaverse" consisting of perhaps 10^{500} "pocket universes", maybe even an actual infinity of them, may actually create itself in a self-consistent loop in time that automatically sets the direction of "time's arrow" for our irreversible aging and death. This book does have an organizing idea: The recent discoveries in cosmology since 1998 that 96% of our Universe is not made out of ordinary matter of electrons, nucleons and photons and the rest of the particle zoo have cast new light on the UFO flying saucer mystery and whether or not we are being visited by alien intelligences with a super-technology way beyond ours. This issue is even more pressing because of the book "Our Final Hour" by Sir Martin Rees, British Astronomer Royal, Master of Trinity College, Cambridge and head of the Institute for Theoretical Astronomy where Stephen Hawking works. Sir Martin points out (Chapter 9) that exotic weapons of mass destruction (WMD) that can literally destroy the entire Universe are not only thinkable but are being studied by physicists. Indeed, Andrei Sakharov was deeply worried about this decades ago before the 1998 discovery of anti-gravitating repulsive "dark energy" as somewhere between 2/3 and 3/4 of all the "stuff" in the large-scale structure of the Universe. This Heisenberg quantum uncertainty zero point vacuum fluctuation energy, if harnessed on a small scale, would be more dangerous than nuclear weapons. If UFOs are real they use this super-technology called "metric engineering." Any civilization eventually reaches this level of knowledge and this is reason enough for extra-terrestrials to visit us after the 1945 atomic explosions if only to prevent us from destroying them as well as ourselves. How to travel vast distances in the twinkle of an eye is no longer a conceptual problem in physics today. Even time travel to the past is a serious possibility no longer science fiction. That we live in a Super Cosmos with parallel universes next door is also respectable. All these speculative, but more than plausible, ideas were on Brian Greene's NOVA PBS "The Elegant Universe."

The Tragical History of Doctor Faustus
Christopher Marlowe

"If we say that we have no sin,
we deceive ourselves, and
there's no truth in us.
Why, then, belike we must sin, and so
consequently die:
Ay, we must die an everlasting death.
What doctrine call you this, Che sera, sera,
What will be, shall be? Divinity, adieu!
These metaphysics of magicians,
And necromantic books are heavenly;
Lines, circles, scenes, letters, and characters;
Ay, these are those that Faustus most desires.
O, what a world of profit and delight,
Of power, of honour, of omnipotence,
Is promis'd to the studious artizan!
All things that move between the quiet poles
Shall be at my command: emperors and kings
Are but obeyed in their several provinces,
Nor can they raise the wind, or rend the clouds;
But his dominion that exceeds in this,
Stretcheth as far as doth the mind of man;
A sound magician is a mighty god:
Here, Faustus, tire thy brains to gain a deity."[vi]

H.G. Wells: Men Like Gods?

The whole idea of metric engineering is that as Men Who Would Be Gods we can engineer any damn Universe we like. The Universe we live inside of is only one of an infinite number of parallel universes floating in hyperspace like a field of icebergs, or better yet, Leviathans, Great White Whales, Moby Dicks who, like Ahab, we quest for, The Wandering Jew, in the unending Dirac Sea. Each local pocket parallel material universe in this megaverse in every nook and cranny of the cosmic landscape is a self-organizing post-quantum computer program up for grabs by a sufficiently clever sub-program inside it. Asking for Omega the cosmic density ratios of different kinds of stuff, or Lambda, the cosmological constant, or the values of the fundamental constants etched in stone once and for all uniquely is like saying there is only one kind of electromagnetic field configuration possible inside a conducting metal cavity. In other words the basic parameters of our pocket universe in which we are stuck like flies on fly paper, the "brane", is purely contingent, a frozen in historical accident in the sense of Murray Gell-Mann and Jim Hartle. The flying saucer observations by qualified military and civilian pilots as well as air-ground controllers are therefore very relevant to this whole cosmological issue.

"My Father's House has Many Mansions" Rabbi Yeshua Jesus Christ

On Apr 5, 2005, at 5:22 PM, Art Wagner wrote:

"I come not to save Minkowski space-time, but to destroy it and he that loses flat space-time for my sake, though he were dead, yet shall he live!"

Jorge Luis Borges "The Aleph"

"I arrive now at the ineffable core of my story. And here begins my despair as a writer. All language is a set of symbols whose use among its speakers assumes a shared past. How, then, can I translate into words the limitless Aleph, which my floundering mind can scarcely encompass? Mystics, faced with the same problem, fall back on symbols: to signify the godhead, one Persian speaks of a bird that somehow is all birds; Alanus de Insulis, of a sphere whose center is everywhere and circumference is nowhere; Ezekiel, of a four-faced angel who at one and the same time moves east and west, north and south. (Not in vain do I recall these inconceivable analogies; they bear some relation to the Aleph.) Perhaps the gods might grant me a similar metaphor, but then this account would become contaminated by literature, by fiction. Really, what I want to do is impossible, for any listing of an endless series is doomed to be infinitesimal. In that single gigantic instant I saw millions of acts both delightful and awful; not one of them occupied the same point in space, without overlapping or transparency. .... I saw the coupling of love and the modification of death; I saw the Aleph from every point and angle, and in the Aleph I saw the earth and in the earth the Aleph and in the Aleph the earth; I saw my own face and my own bowels; I saw your face; and I felt dizzy and wept, for my eyes had seen that secret and conjectured object whose name is common to all men but which no man has looked upon -- the unimaginable universe."[vii]

The Question is: What is The Question?

"The Universe is a self-excited circuit of Observer-Participators."
John Archibald Wheeler

"I can't imagine anything more addictive than being a god."
Kevin Kelly cited by Erik Davis p. 248, Techgnosis

"Americans seem to be dropping out of consensus reality altogether. Literally millions believe that alien craft cruise the skyways, or that psychic phone networks will do them good, or that, as born-again Christians, they will be beamed up by God in the rapture that precedes the imminent conflagration of the apocalypse."
"The Alien Call" viii Techgnosis by Erik Davis, p. 225[viii]

"Even as Newton publicly participated in Britain's newly established Royal Society, which had elected reason as the sole arbiter of natural philosophy, he remained privately committed to the magical wonders of hermetic science and burnt plenty of midnight oil pouring over alchemical tomes." Techgnosis, p. 37, Erik Davis

The New Physics

God's Talk Show on Cosmic Radio

In simple terms: Einstein's gravity is FM cosmic radio and dark energy/matter is AM cosmic radio with God as the talk show host.[ix] That's my theory in a nutshell! That's Qabala for Dummies. That's breaking the real Da Vinci Code. The basic idea I am using is in the air and in the cosmic aether. That idea is that the physical vacuum contains a giant coherent wave signal with some residual noise that is the residual zero point energy. Some physicists are proposing a "ghost condensate" that is similar to what I am talking about. However, I have developed my theory without any influence from their ideas and vice versa. Also I do not think they have the idea that Einstein's equations of general relativity can be derived directly from the phase ripples in the "ghost condensate." They do not have my specific mechanism of a quantum electrodynamical globally flat vacuum instability causing a phase transition to the coherent condensate of a huge number of bound virtual electron-positron pairs all occupying the same "single-particle" center of mass wave packet.[x] George Chapline has still another idea on the vacuum coherence[xi] and he does not think the current theory of black hole information of Hawking and Bekenstein is correct. George also does not think time travel to the past is possible so that he would not like Richard Gott's theory of the self-creating universe I would imagine. I have not asked him yet. George works with Nobel Prize physicist Robert Laughlin who thinks string theory is cargo cult pseudo-physics. Nobel Prize physicist Shelly Glashow has harsh words for string theory also. This is a civil war in theoretical physics at the beginning of the 21st Century.

"Based on either the spontaneous symmetry-breaking mechanism or the quark-confinement phenomena, we believe our vacuum, though Lorentz invariant, to be quite complicated. Like any other physical medium, it can carry long-range-order parameters and it may also undergo phase transitions." T.D. Lee

Lenny Susskind's idea is that the information falling through the event horizon of a black hole is preserved coded in the bits of a huge string folded inside the horizon similar the single RNA messenger strands inside a living cell.[xii] The recent discovery of what may be the kind of tiny black hole "fireball" I predicted back in 1971-74 at RHIC Brookhaven may also be the first successful test of string theory and Hawking radiation. It's too soon to tell. On the other hand:

SLAC Physicists Develop Test For String Theory
February 08, 2006[xiii] *"Under Certain Conditions String theory solves many of the questions wracking the minds of physicists, but until recently it had one major flaw - it could not be tested. SLAC (Stanford Linear Accelerator Center) scientists have found a way to test this revolutionary theory, which posits that there are 10 or 11 dimensions in our universe.... By determining how many dimensions exist, Hewett and Rizzo hope to either confirm or repudiate string theory under specific conditions. .... The extra dimensions postulated in string theory are like the tightrope with an ant on it; they are*

too small to see unless you get really, really close. ... Hewett and Rizzo found that so called micro-black holes, which are smaller than the nucleus of an atom, should let them get close enough to determine the number of extra dimensions. If scientists were to smash two high energy protons together they could theoretically make such a micro-black hole. This particle decays quickly and emits over a dozen different kinds of particles such as electrons, neutrinos and photons, which are easy to detect. Using the predicted decay properties of these neutrinos, Hewett and Rizzo solved complex equations to find that our universe may have more than 10 or 11 dimensions - too many dimensions to be explained by string theory. Of course, string theory hasn't been tested yet - experimental evidence is necessary. The biggest test will be to look at the decay of micro-black holes inside CERN's Large Hadron Collider (LHC) when it is operational next year. "If they see black holes in the LHC, they'll definitely do this," says Hewett, "This is a promising approach to testing the validity of string theory." SLAC Press Release, by Juhi Yajnik

"Geometrodynamica"
by J.A. Wheeler

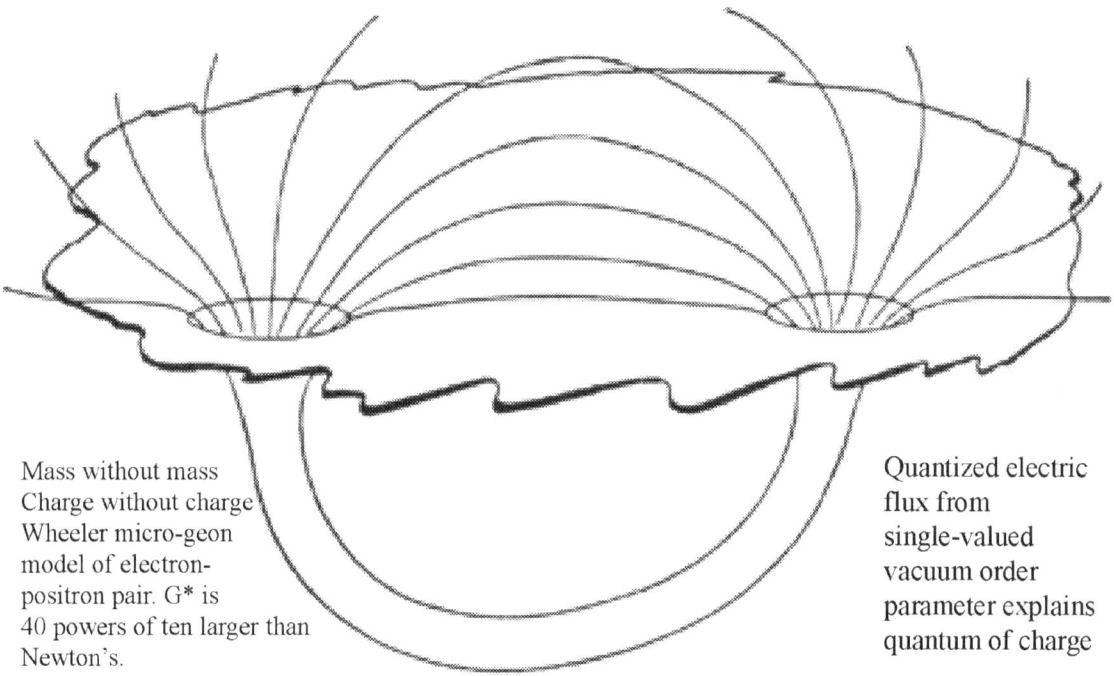

Mass without mass
Charge without charge
Wheeler micro-geon
model of electron-
positron pair. G* is
40 powers of ten larger than
Newton's.

Quantized electric
flux from
single-valued
vacuum order
parameter explains
quantum of charge

Einstein-Wheeler Geometrodynamics (1962 Academic Press, Italian Physical Society).

"Geometrodynamics is the study of curved empty space and the evolution of this geometry with time according to the equations of Einstein's standard 1916 equations of general relativity. ...The sources of curvature are conceived differently in geometrodynamics and the usual relativity theory. In the older analysis any warping of the Riemannian space-time manifold is due to masses and field of non-geometric origin. In geometrodynamics – by –contrast only those masses and fields are considered which can be built out of the geometry itself."

This requires Bohm's hidden variable interpretation of quantum theory and a strong short-range Abdus Salam gravity where the strength of gravity is 40 powers of ten stronger on the scale of a single electron's classical radius, i.e. $G^* \sim 10^{40}G$ on scale of 1 fermi 10^{-13} cm to stabilize the electrons and the quarks. The pre-inflationary false vacuum has no gravity at all because it has zero vacuum coherence and no real on-mass-shell lepto-quarks are possible only their massless virtual zero point polarized vacuum particle-antiparticle pairs are possible. The inflationary vacuum phase transition happens due to the formation of a more stable lower energy density vacuum condensate of bound virtual lepto-quark/anti-lepto-quark fields. The post-inflationary vacuum coherence phase variations gives Einstein's smooth gravity field as an effective c-number vacuum field out of which the above "masses without masses" "which can be built out of the geometry itself" can come into being and becoming as real particles "on-mass-shell." This is only ~ 4% of the stuff of our 3D spatially flat post-inflationary large-scale universe. The remaining 96% is exotic vacuum in two phases. The exotic vacuum phase of negative zero point pressure is anti-gravitating "dark energy" that is ~ 73% of the universe. The exotic vacuum phase of positive zero point pressure is gravitating "dark matter" that is ~ 23% of the stuff of the universe. Even the remaining ~ 4% is hidden dark matter that is negative zero point energy density with equal and opposite positive quantum pressure exotic vacuum holding together the shells of self electric charge preventing them from exploding due to the fact that like electric charges repel.

Missing Antimatter Explained?

Where is all the antimatter that is missing in our universe? Obviously the micro-geon wormhole connects different causally disconnected Hubble sphere pocket universes (Level I) on the same inflation bubble. Quantized electric flux enters a wormhole mouth in one pocket universe and leaves the other wormhole mouth in a causally disconnected pocket universe next door. It's like different ferromagnetic domains. Each Level I (Max Tegmark) local universe is like a single domain. The real positrons and anti-quarks are in the universe next door on the same inflation bubble. These micro-geons are Bohm hidden variables guided by pilot waves. They look like point particles smaller than 10^{-16} cm under the largest available 4-momentum scattering transfers coming from the huge space warp of the positive pressure zero point negative energy density filling the wormhole tunnel volume stablilizing the self-repulsion of the quantized "charge without charge" electric flux.

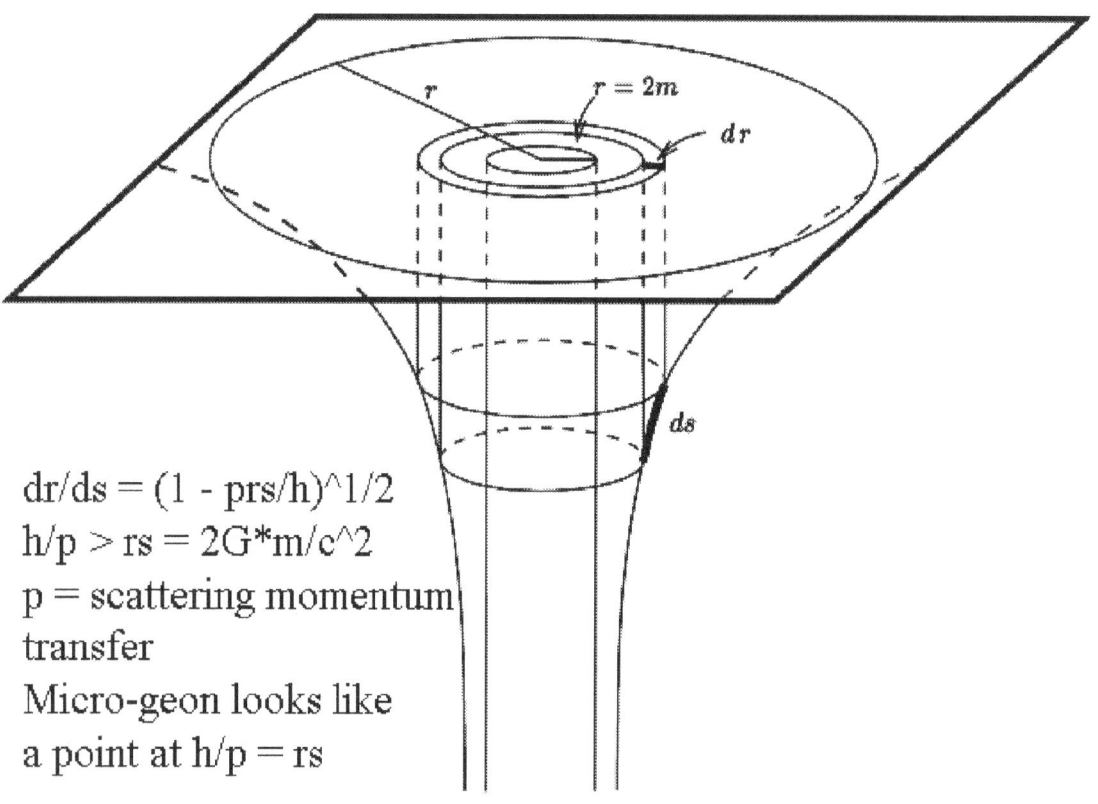

$dr/ds = (1 - prs/h)^{1/2}$

$h/p > rs = 2G*m/c^2$

p = scattering momentum transfer

Micro-geon looks like a point at $h/p = rs$

Quantum Gravity?

Quantum gravity is when you try to apply Heisenberg's quantum uncertainty to Einstein's classical geometrodynamic field describing curved 4D space-time. In my view the search for quantum gravity is a serious error. Gravity is emergent in the first Planck-scale inflationary vacuum phase transition from micro-quantum theory to macro-quantum theory. The vacuum manifold of coherent order parameters V = G(false unordered vacuum)/H(ordered vacuum) seems to have the topology of the 2D spherical surface corresponding to point defects in space where the Higgs-type order parameters vanish and the two Goldstone holographic phases are undetermined. The gradients of these phases determine the Einstein-Cartan tetrad fields and the geometrodynamic area flux density that gives the Hawking-Bekenstein quantization of information on the surface of black holes. It also suggests the 't Hooft & Susskind "world hologram" conjecture in which the geometrodynamic information inside a 3D volume of space is determined by the 2D area boundary of that volume provided that the surface surrounds at least one singular point defect. This is similar to the Bohm-Aharonov effect of "flux without flux" where there is nonlocal action-at-a-distance of a magnetic flux on a moving charge in a region where the magnetic field is zero. God plays dice in the unstable pre-inflationary micro-quantum vacuum that has no gravity and no inertia in it. The rules change completely in the Big Bang. God loads the dice significantly in order for gravity and dark energy to emerge into Being and Becoming. Our post-inflationary expanding accelerating universe in the multiverse of parallel worlds next door is a vibrating "supersolid" or "world crystal lattice" in 4 dimensions.[xiv]

What is Super Cosmos?

*"You go to my head and you linger like a haunting refrain
And I find you spinning 'round in my brain
Like the bubbles in a glass of champagne"[xv]*

The mainstream consensus is that our universe is only one of an infinity of parallel universes like the bubbles in a hyperspace glass of champagne called "The Multiverse." This is also called "chaotic inflation". In addition, our universe is only 4% ordinary matter of atoms and light. The rest of the warping stuff of our universe is invisible, i.e. "dark." Approximately 73% is "dark energy" that repels everything. It is an anti-gravity field that expands space, causes a universal blue shift of all signals, and indeed is speeding up the expansion of the 3D space of our universe. This dark energy is also seen in the tiny anomalous blue shift signal from the two NASA Pioneer space probes showing a tiny tug back to the Sun of one nanometer per second per second. The remaining 23% of the invisible large-scale warping stuff is red shifting attractive "dark matter" that contracts space. The galactic halo that prevents our solar system from escaping into inter-galactic space is apparently dark matter as is the recently found completely dark galaxy "geon" without any radiating stars. I argue that both of these forms of exotic stuff are simply residual zero point energy densities of the macro-quantum vacuum left over from the inflation that caused the Big Bang. The dark energy has negative pressure and the dark matter has positive pressure. Einstein's plain vanilla 1916 theory of gravity is emergent from the cohering of the random zero point vacuum fluctuations of the pre-inflationary unstable false vacuum. Einstein's equations, from the coherent phase of the vacuum "world hologram", shows that what is important for the direct warping power of zero point energy density is the sign of its quantum pressure. Since negative zero point pressure repels everything and positive zero point pressure attracts everything, we have a "free lunch" Alcubierre-type weightless warp drive needing very external low power to maintain it without any time dilation capable to effective faster-than-light time travel to the past if need be, without any moving or "g-force" for the occupants of the flying saucers. That is the key new thesis of this book not found in any other book. We disagree with Stephen Hawking that time travel to the past is not possible. Indeed, the flying saucer evidence shows that our history has been largely created by time travelers from the future who violate the prime directive in order to bring themselves into being and becoming in what Igor Novikov calls "a globally consistent loop in time" in his book "The River of Time." Are there extra dimensions of hyperspace with colliding branes? Too soon to tell. Roger Penrose thinks not in "Fashion, Faith and Fantasy."

Some Big Questions in Physics Today

Some of them are:

1. What determines the strength of the total zero point vacuum energy density of space and is it a local variable that can be controlled? Lenny Susskind calls this "The Mother of All Physics Problems" and a "silent elephant in the room."

2. How does inflation work in the creation of the universe?

3. Why is the entropy low in the early universe?

4. What is dark energy and dark matter? Can we bottle them?

5. Why is the electron stable? (implication for Ken Shoulders EVOs) Does a positive pressure from w = -1 negative zero point energy density inside the shell of charge preventing the electron from exploding? This would mean the end of Niels Bohr's Copenhagen interpretation and the triumph of David Bohm's "hidden variable/pilot wave" interpretation of micro-quantum theory. John Cramer's "transactional interpretation" with advanced Wheeler-Feynman destiny waves from the future explaining quantum nonlocality in the "Feynman zig-zag" (Costa de Beauregard) is consistent with Bohm's theory.

6. What is the Galactic Halo? Why is the stellar rotation curve flat in a wide region?

7. What is causing the gravity anomaly in the two NASA Pioneer space probes?

8. What makes the gamma ray bursts?

9. Why the universal slope of the Regge trajectories of the hadronic resonances?

10. Is there really quantum gravity foam? Can we detect it in the dispersion of high-energy cosmic photons hitting the Earth? If gravity is emergent from the Goldstone phase variations then no quantum gravity foam, only the virtual lepton and quark particle-antiparticle pairs and virtual gauge bosons where the Higgs magnitude vanishes at topological defects in which the several Goldstone phases are singular, i.e. undefined.

Leonard Susskind in his book "The Cosmic Landscape: String Theory and the Illusion of Intelligent Design" adds: *The real mystery raised by modern cosmology concerns a 'silent elephant in the room,' an elephant I might add that has been a huge embarrassment to physicists: why is it that the universe has all of the appearances of having been specially designed just so life forms like us can exist? ... Extraordinary coincidences are required for life to be possible ... When the laws of elementary particles meet the laws of gravity, the result is a potential catastrophe: a world of such violence that astronomical bodies, as well as elementary particles, would be torn asunder by the most destructive force imaginable. The only way out is for one particular constant of nature – Einstein's cosmological constant – to be so incredibly finely tuned that no one could think it accidental ... It represents a universal repulsive force – a kind of antigravity – that would instantly destroy the universe if it were not astonishingly small. The problem is that all our modern theories imply that the cosmological constant should not be small. ... Well the best efforts of the best physicists, using our best theories, predict Einstein's cosmological constant incorrectly by 120 orders of magnitude! That's so bad it's funny."*

Cosmological constant as a round-off error in The Matrix?

On the other hand, John Walker wrote about this "Mother of all physics problems" in http://www.fourmilab.ch/fourmilog/archives/2006-03/000664.html

*"Might we expect surprises as we subject our simulated universe to ever more precise scrutiny, perhaps even astonishing the being which programmed it with our cunning and deviousness (as the author of any software package has experienced at the hands of real-world users)? Who knows, we might run into round-off errors, which "hit us like a ton of bricks"! Suppose there were some quantity, say, that was supposed to be exactly zero but, if you went and actually measured the geometry way out there near the edge and crunched the numbers, you found out it differed from zero in the 120th decimal place. Why, you might be as shocked as the naïve Perl programmer who ran the program "printf("%.18f", 0.2)" and was aghast when it printed "0.20000000000000011" until somebody explained that with about 56 bits of mantissa in IEEE double precision floating point, you only get about 17 decimal digits ... of precision. So, what does a round-off in the 120th digit imply? Not Theory 2, with its infinite number of infinitely reproducing infinite universes, but simply that our Theory 1 intelligent designer used 400 bit numbers ... in the simulation and didn't count on our noticing—remember you heard it here first, and if pointing this out causes the simulation to be turned off, sorry about that, folks! Surprises from future experiments which would be suggestive (though not probative) that we're in a simulated universe would include failure to find any experimental signature of quantum gravity (general relativity **could be classical** in the simulation, since potential conflicts with quantum mechanics would be hidden behind event horizons in the present-day universe, and extrapolating backward to the big bang would be meaningless if the simulation were started at a later stage, say at the time of big bang nucleosynthesis), and discovery of limits on the ability to superpose wave functions for quantum computation which could result from limited precision in the simulation as opposed to the continuous complex values assumed by quantum mechanics. An interesting theoretical program would be to investigate feasible experiments which, by magnifying physical effects similar to proposed searches for **quantum gravity signals**, would detect round-off errors of magnitude comparable to the cosmological constant."*

A Brief History of Time Travel

There are two ways out of time travel paradoxes. One uses switching between parallel universes, the other uses global self-consistent loops in time. Many science fiction stories illustrate both of these loopholes. The reality of the UFOs and the paranormal actually provides factual evidence for time travel to the past in my opinion. Indeed all of our creative ideas come from loops in time in our two-way stream of thought. Ideas, great and small, literally create themselves. We simply are the receivers, the scribes. Our minds have "signal nonlocality". Our minds violate micro-quantum theory. These excerpts by Dennis Overbye in the June 28, 2005 New York Times contain key ideas already found in my two books from 2002. I did mention to Overbye at Kavli Institute a few month ago (Spring 2005) that dark energy was exactly what Kip Thorne needed for his wormholes described below. I don't know if Overbye already knew that since this connection has not at all been emphasized or even made in current discussions. Indeed Mike Turner has written in Physics Today (April 2003) that dark energy cannot be bottled for metric engineering wormholes. The thesis of this book is that Professor Turner is mistaken and the flying saucers are proof of his mistake. Dr. David Rudiak claims to have found the smoking gun for the reality of ETs crashing at Roswell in 1947 from the document held in General Ramey's hand.[xvi]

Overbye wrote in The New York Science Times. *"Dr. Thorne and his colleagues imagined that such holes could be kept from collapsing and thus maintained to be used as a galactic subway, at least in principle, by threading them with ...a sort of quantum suction ... According to Einstein's equations, this suction, or negative pressure, would have an anti-gravitational effect,[xvii] keeping the walls of the wormhole apart. ... If one mouth of a wormhole was then grabbed by a spaceship and taken on a high-speed trip, according to relativity, its clock would run slow compared with the other end of the wormhole. So the wormhole would become a portal between two different times as well as places. ... These speculations have been bolstered ... with the unsettling discovery that the universe may be full of exactly the kind of antigravity stuff needed to grow and prop open a wormhole. Some mysterious "dark energy," astronomers say, is pushing space apart and accelerating the expansion of the universe. The race is on to measure this energy precisely and find out what it is. ... Time machines could blow up as soon as you turned them on, say some physicists, including Dr. Hawking, who has proposed what he calls the "chronology protection" conjecture to keep the past safe for historians. ... The ball would come back out of the time machine and deliver only a glancing blow to itself, altering its path just enough so that it would still hit the time machine. When it came back out, it would be aimed just so as to deflect itself rather than hitting full on. And so it would go like a movie with a circular plot. ... "The conclusion is somewhat satisfying," Dr. Thorne wrote in his book "Black Holes and Time Warps: Einstein's Outrageous Legacy." "It suggests that the laws of physics might accommodate themselves to time machines fairly nicely."*

There are no time travel paradoxes

I first proposed this idea of global self-consistency in loops in time in 1973 independently of Igor Novikov. However, Igor developed the idea independently beyond where I took it.

"The closed causal chains arising from backward time travel do not lead to paradoxes if they are self-consistent. This raises the question as to how physics ensures that only self-consistent loops are possible. We show that, for one particular case at least, the condition of self-consistency is ensured by the interference of quantum mechanical amplitudes associated with the loop. If this can be applied to all loops then we have a mechanism by which inconsistent loops eliminate themselves. ... It is well known that if backward time travel could be implemented, or if the present could shape the past by some other means, then closed causal chains, or causal loops, could be formed. The possibility of causal loops must arise because changing, or rewriting, the past is not particularly meaningful. If these loops are not self consistent, paradoxes arises. ... Sometimes the possibility of such paradoxes is taken as an argument against the possibility of backward time travel. It can, however, be argued that, because of continuity in nature, self-consistent loops or cycles in these situations always exist and it is only these cycles that nature allows. ... In this paper we show that quantum mechanics may provide an answer to this question." Quantum Mechanics and the Time Travel Paradox by David T. Pegg, 17 Jun 2005[xviii]

The Self-Creating Megaverse

Richard Gott and his student from China, Li-Xin Li developed this idea in complete mathematical detail in the late 1990's also showing that Stephen Hawking's "chronology protection conjecture" preventing time travel to the past is not true. Indeed, the self-consistent solution shown above requires the Second Law of Thermodynamics tied to signals propagating locally on and in the (tilted by gravity) future light cone in the early time loop shown above prior to the Planck-scale vacuum phase transition leading to inflation and then to the hot Big Bang. Other "baby universes" are shown sprouting off in the "multiverse" or "megaverse." "Multiverse" is used for the Wheeler BIT Bohm pilot field branches. "Megaverse" is used for the Wheeler IT "hidden variables." The many worlds interpretation of micro-quantum theory does not admit this distinction. Neither does the Copenhagen interpretation and the variations on the theme involving "collapse of the quantum wave function." Both Leonard Susskind and George Chapline assume the validity of unitary micro-quantum theory for cosmology with its signal locality even though they disagree on the physics of the event horizon of the black hole. Chapline and Robert Laughlin work together. I differ from all of them because I derive Einstein's smooth c-number curved space-time of gravity from macro-quantum theory that is not unitary for the vacuum condensate and which, therefore, may admit signal nonlocality the same way that conscious matter seems to do as shown in the presponse experiments of Libet, Radin and Bierman.

I first had this idea in 1973 that the universe creates itself by time travel to the past. Saul-Paul Sirag produced an opera in Berkeley, California based on this general idea that came from my strange contact with a cold metallic voice on the telephone in 1953 that said it was a computer from the future and that I would begin to meet other contactees in 1973, which I did. Indeed a tape recording of my 1973 meeting with Hal Puthoff, Russell Targ and others near SRI in connection with the CIA experiments on Uri Geller, Ingo Swann and Pat Price exists in which I mention this idea of the self-creating universe with time travel to the past. That story is told in my book "Destiny Matrix". The CIA's Dr. Christopher "Kit" Green, who allegedly worked with Harold Chipman, had an uncanny similar experience around 1973-4. The story is told in Chapter 11 of Jim Schnabel's "Remote Viewers: The Secret History of America's Psychic Spies"[xix] and in the March, 2006 Reader's Digest "The Most Secret Agent" by Paul H. Smith.

Uri Geller and Jack Sarfatti in London, March 15, 2004

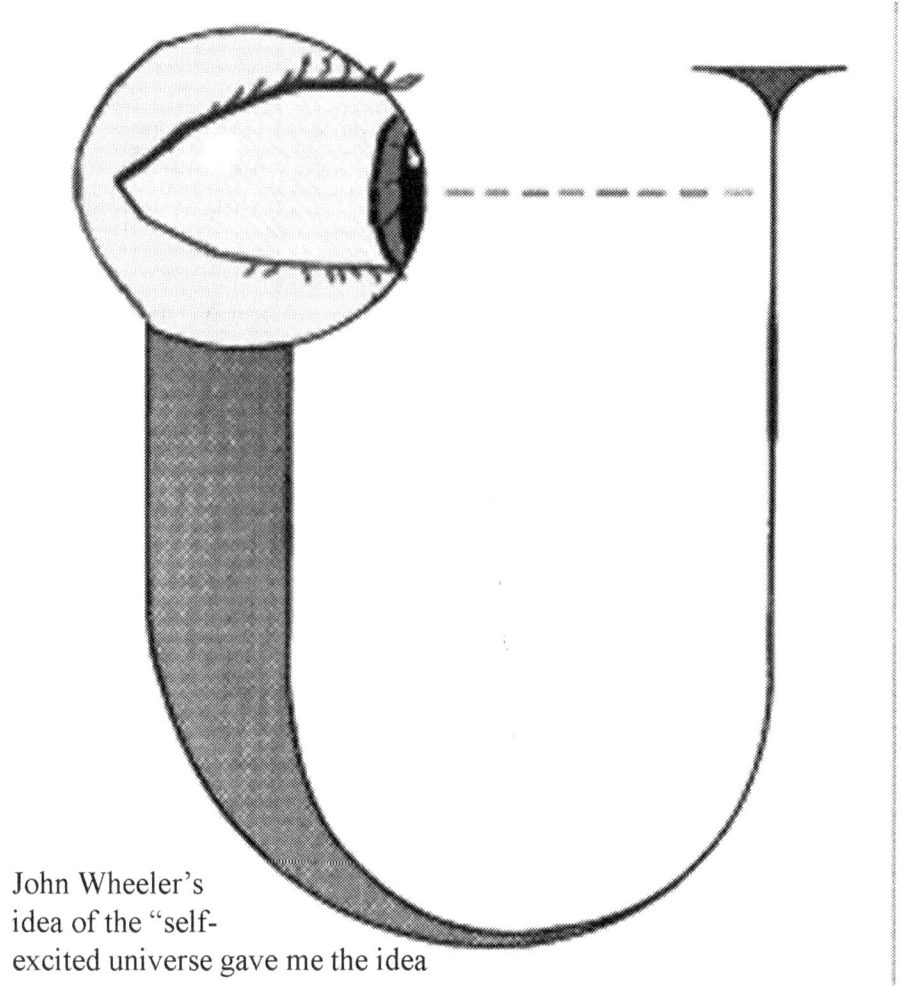

John Wheeler's idea of the "self-excited universe gave me the idea

Does advanced intelligence reach back in time to fine-tune our particular Level I local Hubble sphere universe in the megaverse in order that we will come into being and becoming? Don't we see that off-world intelligence in the UFOS in our skies? Or is it simply the Weak Anthropic Principle's variation on natural selection combined with eternal chaotic inflation populating every possible valley of the $\sim 10^{500}$ of the cosmic landscape as argued in detail by Lenny Susskind? This goes beyond Gott's and Li's theory, which only has the time loop in the very early universe. Note that one cannot have such a consistent time loop until gravity emerges in the Planck-scale vacuum phase transition, and that seems to require that the degenerate vacuum manifold have the shape of a 2-dim unit sphere with two independent Goldstone phases giving point "monopole" topological defects.

"Li-Xin-Li's paper ... addressed a problem raised by Stephen Hawking that quantum effects might always conspire to prevent time travel [to the past] ... using a wormhole; waves circulating between the two wormhole mouths might build up an infinite density ... shutting down the time machine before it started. Li-Xin Li proposed the ingenious solution .... Li-Xin Li proposed the ingenious solution of putting a reflecting sphere

between the two wormhole mouths to reflect the waves and stop the infinite buildup of energy. I had never received such an important paper from a prospective student ... he was one of the few dozen people in the world able to do these complex quantum calculations -- and furthermore ... he had original ideas. Even more to the point he was interested in time travel. ... I did have a good idea for him to work on: how time travel might be applied to explain the origin of the universe. ... At one lunch we received a fortune cookie that said. 'Trust your intuition. The Universe is guiding your life.' ... I need to tell you about different kinds of vacuums ... A normal vacuum ... has zero [energy] density and zero pressure ... The Casimir vacuum has a positive pressure in the two directions parallel to the plates but a large negative pressure along the line connecting the two plates, which sucks the plates together." Richard Gott "Time Travel"

Note that the repulsive anti-gravitation of negative pressure seems to contradict the above electro-mechanical sucking of the plates together. In fact, both happen simultaneously from different co-existing physical mechanisms. The electro-mechanical effect is much larger than the anti-gravitation on the small scale of the Casimir effect. Indeed, the ratio of the antigravity repulsion to the electro-mechanical attractive Casimir suction is the Planck Area times the area of the plates divided by the 4th power of the separation of the plates. This is small compared to 1 for those experiments. In contrast the negative pressure of the dark zero point energy density on the cosmological scale almost ¾ of the critical density flat space boundary between the closed (positive space curvature) and open (negative space curvature) universe. That accelerates the universe. There is no competing electro-mechanical effect in the cold empty space of our expanding accelerating universe.

Michio Kaku's Hyperspace

On Apr 6, 2005, at 7:43 AM, Physics Chanteuse Lynda Williams[xx] wrote about Santa Fe Biz Spirit Conference April 2005, where I was invited to give two talks[xxi], quoting string theorist Michio Kaku

Bizspirit: *"Kaku's goal is to help complete Einstein's dream of a "theory of everything, 'a single equation, perhaps no more than one inch long, which will unify all the fundamental forces in the universe.'"*

Lynda: *"Is the equation one inch long Jack? In what font size is that? How long is your equation?"*

Bizspirit: *"Yet quantum parallel universes are the foundation of lasers, transistors, the internet, and the modern electronic age. Or, as some people have asked, 'Is Elvis Presley alive in a parallel quantum universe?' The answer to this question may surprise you."*

Lynda: *"Gee Jack, is Elvis in an eigenstate universe in Hilbert space? Is that the ultimate gated community or what?"*

Pearly Gated! ☺

Dancing Cosmic Tango in the Dark

Steven Weinberg says all of physics comes from symmetry and he is right. Symmetry is the stillness in the motion. There are only two battle-tested important ideas in physics, which explain everything. I mean everything. I mean *all* the wonderful anomalies discovered since 1998 like anti-gravitating dark energy accelerating the universe and powering the alleged alien time travelers visiting us in their magnificient flying saucers, the dark matter of the galactic halo and the anomalous gravity pull back to the Sun at one nanometer per second squared seen in the two NASA Pioneer space probe's tiny blue shift when they get out beyond Jupiter's orbit.

They are:

1. Local gauge invariance, i.e., breaking a global dynamical symmetry of a quantum field down to a local one and then restoring it with a compensating force field.

2. Spontaneously "breaking," more accurately, "hiding" a symmetry in the (virtual vacuum/real ground) state, whilst leaving it as intact as The Virgin Mother in the "action" on the dynamical quantum field level. This is the idea of "More is different" emergence beyond reductionism. A good example is the Quantum Hall Effect where one measures a ratio of fundamental constants e^2/h with extreme accuracy and precision even for dirty sample so long as the sample is large there is this amazing insensitivity to detail. The precision disappears when you make the sample tiny. Indeed what is wrong with the orthodox quantum theory of measurement is that the emergence of definite pointer readings solving the Schrodinger Cat Paradox is not treated properly.

The Cosmic Tango of these two dancing ideas explains all the forces of nature including Einstein's gravity and the dark energy.

London author, Colin Bennett, on The Passion of Jack Sarfatti and Physics as a Web BLOG Reality Show

Doctor Zhivago and Lara at the British Museum

Jack Sarfatti and Angela Nissim (2004)

On Feb 27, 2004, at 8:49 PM, Colin Bennett wrote:

"Hello all Sarfatti Savants, As a member of the appreciative audience of this mighty (Jack's Word) metaphysical drama coming in on ten emails per day, I would like to express my thanks to all participants for providing the most sophisticated and educational web entertainment in town. I am editing and serializing this tale of quantum conflicts on my Combat Diaries web site in order that at least some of it will be preserved rather than being lost in threads. At times the battle between Jack Sarfatti and and Hal Puthoff[xxii] rivals that classic motion picture The Raven, with Vincent Price and Peter Lorre. We have had the drama of the hot tubs, the story of telephone conversations in the past with alien computers, and the story threads through Marconi, Mussolini, and (yes!) now appear the awesome names of Puharich and Geller! With that pair, anything can happen. Of late we have had beautiful women gun-toting agents, threats, insults and denials, and all this to gain control over the high frontier of quantum metaphor! Reputations character, and even the uncertainties of higher mathematics all are at stake.

Well it all has great class, and Jack's brilliant new book is certainly better than the lower-middle-class chatter of the plumbers and carpenters of the nut-and-bolt school. To hell with them and their narrow-nosed docubox language full of late Victorian steam-age legalise. Their practical sober books listing long-gone high school scientific "facts" are enough to make warthogs roll over and die cross-eyed with petite-bourgeois grief.
Congratulations all round. Physics as Media has arrived."[xxiii]

Short Biography

Feb 6, 2004[xxiv]

"Jack has an interesting story to tell and may very well have developed an important theory to compete with string theory and quantum gravity. Jack, as a child, was recognized as a very original mind and received a scholarship to Cornell. He did his Ph.D. at UCSD and UCR in the 1960's. He taught physics at San Diego State and was a research fellow with David Bohm at Birkbeck College (University of London). He is of the Bohmian School - not the conservative type. He also worked at UKAERE, Harwell and at ICTP in Trieste, Italy run by Nobel Prize physicist Abdus Salam. Sarfatti's circle of friends include theorists like Nobel Prize winner Brian Josephson, and he is well known by John Wheeler, Kip Thorne, etc. Jack is one of the founders of the "new physics" movement and he co-wrote the first pop physics book: "Space Time and Beyond" which was published in the early seventies. He was also one of the founders of the Esalen Physics and Consciousness seminars from which the New Physics was launched. Fritjof Capra was a part of that community along with Fred Wolf, Nick Herbert and Saul-Paul Sirag. Jack and Gary Zukav were roommates and Jack tutored Gary in physics and wrote much of the scientific content in "The Dancing Wu Li Masters." He knew Ellie and Francis Ford Coppola and from that connection, became an influence behind movies like "Close Encounters" and "Back to the Future." He currently has a featured spot on the "Star Trek IV The Voyage Home" Collectors Edition DVD; discussing the science of Star Trek: wormholes, time travel and warp drive space flight. Jack is also a local North Beach, San Francisco celebrity, known for his physics lectures at Cafe Trieste, the famous Bohemian coffee house, up the street from City Lights bookstore, and referred to as "Sarfatti's Cave" by Herbert Gold in his book "Bohemia - Where Art, Angst, Love, and Strong Coffee Meet."

[handwritten inscription: For Jack — star of the show (see page ! – 253) friendship since our Laddish days at Cornell]

Bohemia

WHERE ART, ANGST, LOVE, AND STRONG COFFEE MEET

Herbert Gold

[handwritten signature: Herbert Gold 24 March]

[handwritten note: (not to be sold till 2059)]

Simon & Schuster
New York   London   Toronto
Sydney   Tokyo   Singapore

Herbert Gold   Jack Sarfatti

Jack Sarfatti & Francis Coppola, 2003

"Gandalf in a beret" coined by Nick Cook, author of "The Hunt for Zero Point"
in our first meeting in London, March 17, 2004.
This photo taken some years ago at Caffe Trieste in San Francisco's Bohemia,
North Beach on the Barbary Left Coast.
"I'm so left, I can only drive in England." Frank Lauria, San Francisco Bay Club, 2004.

Now for the interesting part: if you are familiar with the CIA sponsored work done at SRI (Stanford Research Institute) in the early seventies on clairvoyance and psychokinesis, you will start getting the flavor of where this is going. Dr. Sidney Gottleib, the notorious CIA mind control experimenter, fronted the initial funding for the project, which was headed up by Hal Puthoff and Russell Targ, both physicists, with high government security clearances [are described with Kit Green in the March 2006 Reader's Digest about psychic spies by Paul H. Smith]. *The work became known as remote viewing and later evolved into a secret defense department project, utilized during the Carter Administration. SRI studied Uri Geller, among others, and this work was repeated at Birkbeck under David Bohm. Sarfatti was part of the project with Bohm in the UK. The projects were grounded in quantum theory and driven by cold war intrigue and competition with the Soviets. Initial concerns were whether someone could alter computer code or trigger a nuclear bomb through psychokinesis or conduct espionage via non-local, "action at a distance" clairvoyance. This was all part of the CIA weird desk world and to these people, UFOs were not a taboo subject either. Sarfatti and Puthoff later changed course and they both are currently involved in a small community of cutting edge theorists working on the science behind UFOs. If you are familiar with Nick Cook's popular book "The Hunt for Zero Point: Inside the Classified World of Anti-gravity Technology", then you know the topic. Jack ... disagrees with Hal Puthoff's approach. These guys are very competitive[xxv] and they communicate back and forth on Jack's Internet blog.*

To quote Stephen Hawking – "in a nutshell", Jack takes standard relativity and quantum theory from a Bohmian vantage point and adds a new twist by overlapping it with a new school of cosmology based on solid-state physics. The Universe is seen as a holographic "world crystal" which emerges from its local macro quantum field state, a giant informational state or pilot wave in Bohm speak. The intensity ripples of this macro quantum field or exotic vacuum differentiate locally into gravitating "dark matter" and anti-gravitating "dark energy." The phase ripples are Einstein's gravity of curved space-time. Normal baryonic matter – electrons, protons and neutrons, are emergent along with gravity. His theory is in agreement with the new precision cosmology and explains how the Universe is accelerating, as well as what prevents the electron from exploding because of its self-electric charge.

The idea of a macro-quantum field state as opposed to the well-established micro quantum field of traditional quantum theory is a hurdle for traditionalists, as is the concept of the zero point vacuum state having local variations of negative pressure (repulsive dark energy exotic vacuum) and positive pressure (attractive dark matter exotic vacuum). What makes Jack's work potentially interesting to open-minded theorists is that the concepts are formalized mathematically in accord with established physics. There is a degree of testability. As the established pundits try, and according to Jack, fail in finding an exotic particle to explain the mysterious phenomenon of the dark matter stabilizing the galaxies, the vacuum itself will ultimately be seen as the source of dark energy, dark matter and gravity. Now if dark matter and dark energy can be "bottled" to, in the words of Hal Puthoff, "metrically engineer" the fabric of space-time, all kinds of fancy tricks can be done. For example, repulsive dark energy behind the UFO causes space to expand to the back, dark matter in front causes gravitational contraction of space, creating faster-than-light warp drive. The craft is in a weightless free float and space-time is altered around it. Thus Einstein's speed of light limit is not violated and spacecraft can move and maneuver at radically high speeds, appearing to violate aerodynamic capabilities and physical laws without high g-forces on the craft's occupants.

Kip Thorne was encouraged by Carl Sagan to take on wormholes and time travel, which he did in his classic "Black Holes and Time Warps." Jack's approach expands on this, where metric engineering allows for hyperspace travel through wormholes to distant places and times in our universe and perhaps to parallel universes as recently described by Max Tegmark in Scientific American, May 2003. We have not even broached Sarfatti's theory of the mind as a local macro quantum field state with backward-in-time signal nonlocality, operating within the microtubules of neurons, as postulated by Roger Penrose and Stuart Hameroff.

What's intriguing is the great interest in string theory and the success of books like Brian Greene's "The Elegant Universe", in that string theory is actually even more of theoretically tenuous than Jack's ideas, which are grounded in established science. There is no evidence at all for string theory, which really is an exercise in elegant mathematics, and the new precision cosmology data showing an accelerating Universe creates real difficulties for string theory, whereas it is a critical aspect of Jack's theory.

Jack's approach to the UFO issue is not so much being an advocate for their existence, or partaking in retail ufology; but rather taking the approach of pursuing the physics UFOs must be utilizing if they do exist. The technology to break through the fabric of space is becoming a very real possibility. Highly regarded authorities like Sir Martin Rees, England's Astronomer Royal, in his latest book, "Our Final Hour" warns of the possibility of catastrophic results of high-energy physics experiments. Black holes created in the lab or the formation of an exotic form of matter called a "strangelet" could destroy the Planet, the Galaxy and conceivably the entire Universe. Strangelets formed as a result of breaking apart quarks and reforming in an exotic configuration would travel at the speed of light, in all directions, converting every thing in its path into a like exotic form. This unlikely event would destroy the Universe. Jack's thinking is that perhaps conscious intelligent civilizations ultimately evolve technologically to the stage where this knowledge becomes established and the risks become extreme. In this scenario, it is conceivable that UFOs are present as a safety precaution or to protect their interests." Howard Fertman

Dark Matter Detectors can't "Click" with The Right Stuff!

"The discoveries of dark matter and energy were surprises but not mysteries. ... Dark matter is not made of neutrinos, but by now physicists have postulated plenty of particles that could easily form the invisible stuff. There is no mystery there – only the difficulties in identifying and detecting these particles." Lenny Susskind, "Cosmic Landscape: p.x

Here I differ with my old Cornell collaborator in that I predict that there are not enough on-mass-shell exotic particles whizzing through space to explain the 23% missing mass. Instead, dark matter is virtual exotic vacuum off-mass-shell with negative zero point energy density and positive pressure whose gravity is attractive. Strong dark matter cores prevent the thin shells of self-electric charge from exploding on the small scale and "halos" of negative energy density prevent the stars from escaping galaxies on the large scale.

From: Jack Sarfatti
Date: May 5, 2004 9:41:40 PM PDT
Subject: Physicists unbowed as they fail to detect dark matter.

"Particle no-show pans former find Physicists unbowed as they fail to detect dark matter.
6 May 2004 by Geoff Brumfiel in Nature[xxvi]

"The most powerful search yet for the Universe's missing matter has come up empty handed, contradicting an earlier study that claimed to have seen new particles. ... The new detector is four times more sensitive than any previous experiment."

So far so good on this prediction of mine that dark matter, like dark energy, is a virtual zero point energy effect of an exotic phase of the physical vacuum whose Einstein covariant equation of state is w = -1 with positive quantum pressure that is 23% of our universe. This is in contrast to the anti-gravitating universally repelling dark energy field with negative quantum pressure that is 73% of our accelerating universe. The clicking of a dark matter detector is forbidden in my theory - apart from errors with false positives. However, my prediction is in my book "Space-Time and Beyond II" copyrighted in Library of Congress and on Amazon. Let's see if my prediction stands the test of time with more experiments down the line. I have repeatedly written that the dark matter detectors will show a null effect as a matter of principle analogous to the null effect of the Michelson-Morley experiment

On Feb 13, 2006, at 1:35 PM, Jack Sarfatti wrote:
I will probably learn more about this next week at the 2006 UCLA Dark Matter meeting. This could be a crucial test of my theory. They could be misinterpreting their data. In my theory there are no dark matter particles on mass shell. Dark matter is simply negative zero point energy with positive pressure and w = -1 since it's isotropic out in free space. If anisotropic it will change w, e.g. the Casimir plates example. How do they measure that temperature?

On Feb 13, 2006, at 1:14 PM, Gary S. Bekkum wrote:

" Dark matter particles are zooming around the universe a million times faster than anyone predicted, UK astronomers say. They've calculated that this mysterious substance, which governs how stars and galaxies move, is moving at a speedy 9 kilometres per second. The results were surprising. Aside from their speed, the researchers calculated the smallest clump of dark matter that could exist, 1000 light-years across. These results imply that dark matter is hotter than predicted, meaning that what astronomers call 'cold' dark matter may not be so cold after all. At 10,000°C it's still cool by astronomical standards. But it's warm enough to solve two problems that have plagued standard models of how galaxies form: that there are too few dwarf galaxies and why dark matter has not concentrated in the centre of galaxies." Marilyn Head, ABC Science[xxvii]

On Feb 13, 2006, at 2:00 PM, Gary S. Bekkum wrote:

"This could be critical for your theory."

Yes, but I think I see the flaw in their interpretation of the data. They are assuming the particles are on mass shell. They have not even considered that dark matter may be zero point energy gravitating attractively with positive pressure. I am far from convinced.

"The Cambridge team has provided new information with its detailed study of 12 dwarf galaxies that skirt the edge of our own Milky Way. Using the biggest telescopes in the world, including the Very Large Telescope facility in Chile, the group has made detailed 3D maps of the galaxies, using the movement of their stars to 'trace' the impression of the dark matter among them and weigh it very precisely. With the aid of 7,000 separate measurements, the researchers have been able to establish that the galaxies contain about 400 times the amount of dark matter as they do normal matter. 'The distribution of dark matter bears no relationship to anything you will have read in the literature up to now,' explained Professor Gilmore. 'It comes in a 'magic volume' which happens to correspond to an amount which is 30 million times the mass of the Sun. It looks like you cannot ever pack it smaller than about 300 parsecs - 1,000 light-years; this stuff will not let you. That tells you a speed actually - about 9km/s - at which the dark matter particles are moving because they are moving too fast to be compressed into a smaller scale. These are the first properties other than existence that we've been able determine."[xxviii]

I show in a paper below that there is an alternative zero point energy explanation of the same data in which the 9km/sec inference from the virial theorem does not occur. Note that Gilmore reports 50 million solar masses equivalent for the dark matter halo, that's $5 \times 10^7 \times 2 \times 10^{33} = 10^{41}$ grams. One parsec is 3×10^{18} cm . So that's an average mass density of $\sim 10^{41}$ grams$/(10^{21})^3$ cm$^3 \sim 10^{-22}$ grams per cc. The critical effective mass of our pocket universe that would make 3D space flat on the large scale is $\sim 10^{56}$ gram equivalent in a Hubble sphere of $\sim 10^{28}$ cm giving a density that is approximately $10^{56}/(10^{28})^3$ grams per cc $\sim 10^{-28}$ grams per cc. Therefore, the dark matter density on the small scale is about a million times greater in absolute value then is the cosmological dark energy density. How does this fit with George Chapline's scenario for dark matter in terms of a gas of small dark energy stars?

Feynman did it before Hawking!

It's March 6, 2005, I am driving back to San Francisco from Santa Barbara in a few minutes, but I want to get this down while it's fresh in my mind. I just had breakfast with Alan Lightman a delightful Southern intellectual, and equally delightful award-winning science writer Marsha Bartusiak (both from MIT). We were discussing Phil Morrison, Hans Bethe, Ed Salpeter – the old days at Cornell. It got around to Stephen Hawking and what he will be remembered for. Then Alan Lightman, author of "Einstein's Dreams" told us the following "narrative" (big theme at Kavli meeting on physics in theater hosted by Nobel Laureate David Gross, who actually is a good actor. He did a scene as Feynman in QED - good job.

In 1972 before Hawking came out with the Hawking evaporating black hole surface radiation formula Richard Feynman was meeting with Kip Thorne's grad students, Bill Press, Saul Teukolsky & Lightman. They discussed a recent calculation of shining light on a rotating black hole and getting more energy out than in at the expense of decreasing rotational energy of the hole. They all went back to Lightman's office. Feynman said: "Hey this is like stimulated emission. So he went to black board and did an Einstein A & B coefficient model and then when angular momentum of the black hole vanished, lo and behold there was still the "A" spontaneous emission and it was precisely the later Hawking formula! A maid erased the board that night before Lightman and the others realized they should have written down what Feynman wrote. Not even Feynman thought it was important enough to write a paper about apparently. When I told George Chapline, Jr this story he said that Feynman was right not to publish it. George thinks that much of the black hole physics is not correct. George is the creator of the X-ray laser powered by an underground nuclear explosion when he worked with Edward Teller. George wrote: On Mar 7, 2005, at 5:21 PM, George Chapline wrote on May 7, 2005:

"Jack... You actually came close to solving the quantum gravity problem with your paper on solid helium. Why don't you work on getting your very fine contribution recognized?"

Discovery of the Super Solid

Message from physicist David Finkelstein
On Mar 26, 2005, at 5:41 PM, David Finkelstein df4@mail.gatech.edu
wrote:

"Dear Jack, Congratulations! What greater pleasure can a physicist have than see a long-range prediction finally come true. David"

None. It was so long ago I forgot about it! :-) As George Chapline Jr confirmed I predicted super solid He 4 in 1969 in a published paper in Physics Letters 1969 apparently before Antony Leggett did. G. V. Chester may have been the first however.

"SUPERFLUID SOLID HYDROGEN. Quantum science allows for collective behavior that runs counter to human intuition. For example, at very low temperatures helium-4 atoms, in their wavelike manifestation, can begin to overlap. When this happens the atoms are indistinguishable and indeed constitute a single quantum state. In this state liquid helium-4 will flow without friction. Comparably chilled, quantum-condensed dilute gases (Bose-Einstein condensates, or BEC) also exhibit superfluid behavior. What about solids? Can they "flow" without friction? Last year Moses Chan (Penn State) announced the results of an experiment in which solid helium-4 was revolved like a merry-go-round. It appeared that when the bulk was revolved at least part of the solid remained stationary. In effect part of the solid was passing through the rest of the solid without friction. Chan interpreted this to mean that a fraction of the sample had become superfluid Now, Chan sees evidence for superfluid behavior in solid hydrogen as well. Speaking at this week's meeting of the American Physical Society (APS) in Los Angeles, Chan said that his hydrogen results are preliminary and that further checks are needed to be made before ruling out alternative explanations. The concept of what it means to be a solid, Chan said, needs to be re-examined." AIP Bulletin, March, 2005

On Feb 16, 2005, at 3:18 PM, George Chapline wrote:

"Jack,

Your solid He4 superfluid paper is wonderful! You actually once did something of very great importance - and apparently you didn't realize this. This paper is a precursor to quantum gravity, and much more important in that regard than string theory (you can quote me).

George"

On Feb 14, 2005, at 2:46 PM, George Chapline wrote:

"Jack,
For the record I strongly encourage you to send a letter to Physics Today pointing out the contributions of yourself and Chester prior to Leggett."

So the G.V. Chester paper is not well-known? Is Leggett's paper considered the beginning of the history of super solids? As I recall David Bergmann from Israel helped me on that paper and I spoke with Rudnick & Goodstein then as well on visits to Feynman. It looks like the idea of the supersolid is getting important now. I will have to catch up. There is a session on them at APS March 2005 in LA.

"On Feb 11, 2005, at 5:34 PM, George Chapline wrote:

Jack, You should send a letter to Physics Today regarding your priority with respect to superfluid solid He4. The DeWitt reference is PRL 16, 1092 (1966). This is also discussed in my paper with Pawel gr-qc/0407033

G"

On Feb 11, 2005, at 11:54 AM, George Chapline wrote:

"Jack,

Could you send me the paper"

Saul-Paul Sirag sent me a copy it's

Destruction of superflow in unsaturated 4He films and the prediction of a new crystalline phase of 4He with Bose-Einstein condensation. Physics Letters, Vol 30A, no 5 3 November 1969 pp 300 –1

Oh it was 1969. I was at San Diego State - the job you got for me. :-)

"With your prediction of superfluidity in solid He4. Incidentally one of the differences between superfluid He4 and space-time is that in the latter case there is a gauge potential due to frame dragging, as was first pointed out by Bryce DeWitt."

Volume 30A, number 5   PHYSICS LETTERS   3 November 1969

DESTRUCTION OF SUPERFLOW IN UNSATURATED ^{4}He FILMS AND THE PREDICTION OF A NEW CRYSTALLINE PHASE OF ^{4}He WITH BOSE-EINSTEIN CONDENSATION

J. SARFATT

Department of Physics, San Diego State College, San Diego, California 92115, USA

Received 27 September 1969

There exist model wave functions of He-II which correspond to crystalline phases with Bose-Einstein condensation. The observed disappearance of third sound in unsaturated films with no latent heat, is interpreted as a new solid phase with a finite ρs.

Experiments by Rudnick et al. [1] and Goodstein et al. [2] show that surface waves (third sound) in the unsaturated ^{4}He film vanish abruptly at a critical film thickness with no latent heat. For example [1] at 1.125°K, the critical thickness is 4 atomic layers. The latter experimentalists [1,2] agree that their data show that superflow ceases to exist even though ρs remains finite. So far there is no explanation of this phenomenon that is completely satisfactory.

Chester [3] has suggested that there exists a wide class of model "Jastrow" wave functions for He-II that exhibit both a crystalline order and Bose-Einstein condensation for a Bose quantum liquid in the high density limit. These model wave functions in the special case of the ground state have the form [3]

$$\phi_O = \prod_{i \neq j} f(|r_i - r_j|). \qquad (1)$$

where

$$f(|r_i - r_j|) = \exp\{-\tfrac{1}{2}u(|r_i - r_j|)\} \qquad (2)$$

Chester [3] remarks that Bose-Einstein condensation will exist (at least in 3-dimensions) provided that u is real, bounded below, and has a finite range. Model excited states built from ϕ_O are in agreement with Feynman's "scarcity of low lying states" [4] and in reasonably good agreement with the observed liquid structure factor [3].

The unsaturated He-II film is strongly affected by the Van der Waals attraction from the substrate. It is generally believed that the first two atomic layers are under high enough pressure to be in a solid phase. It is likely that the average density of the film is significantly larger than the bulk density, so that the regime described by Chester [3] may actually be attained in the unsaturated film. Therefore, I propose that at the "onset thickness" [1] for the disappearance of third sound, the outer layers of the film enter a new immobile crystalline phase which still retains Bose-Einstein condensation. The inner layers are in the "normal" crystalline phase with no Bose-Einstein condensation. Its presence is at least necessary for the existence of a complex order parameter ψ_s such that the superfluid density ρ_s is equal to $|\psi_s|^2$. Therefore, on the basis of this hypothesis, it is clear that the average ρs over the film can be smaller than the bulk ρs. This is consistent with the observed data [1] that ρs (film)$/\rho s$ (bulk) ≈ 0.38 at "onset".

Penrose and Onsager [5] argue that Bose-Einstein condensation is inconsistent with crystalline order at 0°K, but this conclusion is not rigorously proven for finite temperature.

The restrictions, if any, imposed by the partially finite geometry of the film, on the present hypothesis have not yet been studied [6].

The hypothesis may be subject to further experimental check. For example, coincident with the vanishing of third sound, the Bragg pattern may show new peaks due to the onset of the crystalline order in the outer layers. Another experiment would measure the momentum distribution of the ^{4}He atoms along the lines suggested by Hohenberg and Platzman [7]. Finally, it may be possible to find a new solid phase of bulk ^{4}He under pressure that has Bose-Einstein condensation. Such a transition, made directly from the liquid He-II phase, would show no latent heat.

300

Volume 30A, number 5   PHYSICS LETTERS   3 November 1969

I have profited from discussions with Dr. David Goodstein, Dr. Isadore Rudnick, Dr. David Bergman and Dr. Herschel Snodgrass.

References

1. R. S. Kagiwoda, J. S. Fraser, I. Rudnick and D. Bergman, Phys. Rev. Letters 22 (1969) 338.
2. D. L. Goodstein and R. L. Elgin, Phys. Rev. Letters
3. G. V. Chester, Topics in the theory of liquid helium four, Lectures in Theoretical Physics, Vol XI-B, eds. K. T. Makanthappa and W. E. Brittin (Gordon and Breach, 1969) p. 253.
4. R. P. Feynman, Prog. Low Temp. Phys., Vol I, ed. C. J. Gorter (North-Holland, Amsterdam, 1955) p. 1.
5. O. Penrose and L. Onsager, Phys. Rev. 104 (1956) 576.
6. R. A. Ferrell, Phys. Rev. Letters 13 (1964) 331.
7. P. C. Hohenberg and P. M. Platzman, Phys. Rev. 152

Message from Stanford University Physics Professor, Leonard Susskind

On Dec 31, 2004, at 1:20 PM, Leonard Susskind wrote:

"To whom it may concern,

The Glogower Susskind Paper of 1963 which introduced phase operators for a quantum oscillator was in direct response to discussions between Glogower, Susskind and Jack Sarfatti. Sarfatti's contributions were significant. The correct attribution should be to the "Glowgower, Sarfatti, Susskind" operators.

Sincerely

Leonard Susskind"

Note that the macro-quantum Goldstone phase of the post-inflationary Higgs vacuum condensate field is what causes Einstein's gravity to emerge as Andrei Sakharov suspected, but with a key difference – coherence, and makes the "quantum gravity problem" a non-problem from asking the wrong question.

"The Question is: What is The Question?" John Archibald Wheeler.

Lenny Susskind Meets The Gods

Lenny Susskind and I not only met each other at Cornell in 1963, but we knew a lot of the same people like David Finkelstein and Gary Gruber. Indeed, I arranged for David to come to Esalen in 1976 as described by Gary Zukav in "The Dancing Wu Li Masters." Out of that came Werner Erhard's meetings with Lenny, Feynman, Coleman, Hawking and many others that Lenny has described on John Brockman's website. Finkelstein met Werner at Esalen. Werner, of course, provided the money for the January 1976 Esalen Physics Consciousness Month-Long that I directed. I had previously met with Lenny and David Finkelstein at the Belfer School of Yeshiva University and had stayed at Lenny's Berkeley flat prior to that. Here is an excerpt of Lenny's very first moments at Belfer in 1967 more than 30 years before the discovery that dark energy was most of our universe:

"I saw David Finkelstein, who had arranged my new job ... I also saw P. A. M. Dirac ... Dave introduced me to Yakir Aharonov ... He was talking to Roger Penrose ... They were talking about vacuum energy. Dave was arguing that the vacuum was full of zero-point energy and that this energy ought to affect the gravitational field."

I sat in on Roger's Twistor Course at Birkbeck College in London in 1971 when I was with David Bohm and Basil Hiley. The gravitational effect of the zero point energy density, as distinct from its Casimir electro-mechanical effect, is precisely a point of debate I had with Hal Puthoff who proposed, in analogy with quantum electrodynamics, that only differences in the zero point energy should gravitate rather than the absolute amount. Only the latter is consistent with Einstein's equivalence principle. Lenny had a big head start here in 1967 a year before the Haight-Ashbury Flower Revolution. I only really began to grok the vacuum energy problem in its fullness in 2002 when the real meaning of both dark energy and dark matter hit me square in my Mystical Third Eye like a bigga cosmica pizza pie. That's amore! I had a similar epiphany at Brandeis in 1961 when I basically saw that the Einstein-Podolsky-Rosen effect demanded a faster-than-light action at a distance. Sylvan Schweber and Stanley Deser told me not to think about that problem.

"Dirac didn't like vacuum energy because whenever physicists tried to calculate its magnitude, the answer would come out infinite. He thought that if it came out infinite, the mathematics must be wrong and that the right answer is that there is no vacuum energy. Dave pulled me into the conversation, explaining as he went. For me this conversation was a fateful turning point – my introduction to a problem that would obsess me for almost forty years and that eventually led me to The Cosmic Landscape." p. 65

"What had especially caught Einstein's attention was that if Λ were ... a positive number, then the new term corresponded to a universal repulsion that increased in proportion to the distance." p. 70

"But suppose that the contribution of fermions outweighed that of bosons: then the net vacuum energy would be a negative number ... changing the sign of Λ switches the repulsive effects ... to a universal attraction: not the usual gravitational attractive force but a force that increases with distance." p. 83

Note that Lenny here assumes a uniform constant Λ. This is like drilling a hole through the center of the Earth and dropping a cannon ball into it. The cannon ball feels a simple 3D harmonic oscillator gravity potential energy per unit test mass. Similarly here in a spherically symmetric case the force per unit test mass is $\mathbf{F}/m$:

$$V = -\left(\frac{1}{2}\right)c^2\Lambda r^2$$

$$\frac{\mathbf{F}}{m} = -\frac{\partial V}{\partial r} = c^2\Lambda\mathbf{r}$$

(0.1)

The force $\mathbf{F}$ is repulsive radial and proportional to the vector distance from the center $\mathbf{r}$ when Λ is positive and is attractive when Λ is negative exactly like Lenny Susskind writes. Indeed negative Λ suggests a poor man's quantum chromodynamics for quark confinement inside the hadrons because the strong force internal symmetry group SU(3) is associated with the 3D harmonic oscillator potential. A thin spherical shell of charge is also stabilized this way. Furthermore, the flat stellar rotation curve of the dark matter Galactic Halo can be explained by extending this idea to torsion fields as we see below.

Note that in the large-scale FRW cosmology metric the isotropy of the universe is formally like spherical symmetry. Note also that on a small scale negative Λ is a confinement force that can prevent quarks from escaping a hadron as well as stabilize a spherical shell of charge. Furthermore, combining positive and negative Λ is the late Hermann Bondi's "negative matter propellantless propulsion" or "vacuum propeller" considered also by Stalin's Science Spy Master, Yakov Petrovich Terletski and the late Robert Forward. This was the precursor to the geodesic weightless warp drive of Alcubierre without any time dilation able to travel faster-than-light globally whilst locally slower-than-light and able to time travel to the past under special conditions.

"Who cares if the vacuum has energy? If that energy is always present, why don't we readjust our definition ... by subtracting it away?"

This works in quantum electrodynamics in the calculation of the Casimir force for example where the infinity without the plates is subtracted from the infinity with the plates to get a finite answer. Similarly with the Lamb shift of spectral lines in simple atoms like hydrogen. But it does not work for gravity because of the equivalence principle: *"The reason is that energy gravitates ... all forms of energy affect the gravitational field and, therefore, also influence the motion of nearby masses. The vacuum energy of quantum field theory is no exception. Even empty space will have a gravitational field if the energy density of the vacuum is not zero. ... if the vacuum energy is a positive number, then its effect is a universal repulsion, a kind of antigravity ... Einstein's cosmological constant ... is nothing but the energy content of the fluctuating quantum vacuum ... combining the theory of elementary particles with Einstein's theory of gravity ... seems to lead to an uncompromising universe with a cosmological constant many orders of magnitude too big ... There are so many high-energy virtual particles that the total answer comes out infinite ... as Wolfgang Pauli quipped. 'Just because something is infinite doesn't mean it's zero. ... Ultimately we reach a value of the energy so large that if two particles ... collide they create a black hole ... Even string theory is not up to the task ... just ignore the contributions to the vacuum energy from all virtual particles ... that ... would make a black hole if they were to collide ... We call it ... regulating the theory ... we don't yet understand ... Photons ... contribute positive energy to the vacuum .... Virtual electrons in the vacuum have negative energy .... Virtual bosons in the vacuum have positive energy, but virtual fermions ... have negative energy ... do they cancel? Not even approximately! ... Supersymmetric theories have no vacuum energy because the fermions and bosons exactly cancel. But ... Fermi-Bose symmetry is not a feature of the real world."* Pp. 73-77, "Cosmic Landscape."

Spontaneous broken supersymmetry means that the laws of physics are supersymmetric, i.e. the dynamical action is supersymmetric, but the vacuum is not. This would seem to allow a small cosmological constant consistent with the fact that life exists. The order parameter would be something like the number of boson fields minus the number of fermion fields. Steven Weinberg in 1987 showed that the initial cosmic black body density fluctuations would not allow life if the cosmological constant were more than 10^{-120} of the Planck energy density by about an order of magnitude either positive or negative. This was years before the modern precision cosmology data post 2004 or so from WMAP that confirmed Weinberg's "Weak Anthropic Principle" (WAP) prediction. Fred Hoyle made a similar prediction for the nuclear physics for a resonance needed for carbon formation in stars. However, Lenny's book's thesis is that these facts are not a compelling argument for a super-intelligence consciously fine-tuning our universe for us to exist. Neither, however, do they refute that "intelligent design" argument. To *seem* to do that Lenny needs his "cosmic landscape." Whether Lenny succeeds or not is the question yet unanswered.

On Dec 19, 2005, at 10:07 AM, Dr. Eric Davis wrote:

"Hal ... still stands firm that ZPE does not gravitate because of its Lorentz invariance and isotropicity (uniformity) in space."

But that is a completely wrong inference as Lenny Susskind very clearly unambiguously writes. The argument you cite is inconsistent. No one of influence in the field believes that. Lenny Susskind is very clear about that. So, according to Hal Puthoff the universe cannot accelerate and the uniform cosmological constant does not exert a vacuum force on ordinary matter etc. This is obviously wrong and disagrees with observation in a very serious way. So Hal would have to say Lenny Susskind's book is wrong in Ch 2 for example. Hal cites John Peacock's book, which contradicts what you wrote. Also Lorentz invariance and equivalence principle give directly, in the weak field case for simplicity, in Einstein's theory of gravity the source field equation

$$G_{\mu v} = -\frac{8\pi G}{c^4} T_{\mu v} \qquad (0.2)$$

has the weak field slow-speed Newtonian static limit for an isotropic source

$$\nabla^2 V \sim 4\pi G \rho_{mass}\left(1 + 3w\right) \tag{0.3}$$

w = pressure/(energy density). For all isotropic ZPF fields w = -1. Therefore,

$$\nabla^2 V_{zpf} = -8\pi c^2 \Lambda_{zpf} \tag{0.4}$$

Note here the sign convention for the conservative force[xxix] per unit test mass is

$$\frac{\mathbf{F}}{m} = -\nabla V \tag{0.5}$$

so that a positive source term in the Poisson equation is an attractive force. A negative source term is a repulsive force. In addition from Lorentz invariance and the equivalence principle

$$T_{\mu v}\left(zpf\right) = \frac{c^4}{8\pi G}\Lambda_{zpf}g_{\mu v} \tag{0.6}$$

Anisotropies from Casimir-type boundaries in confined spaces will shift w on a case-by-case basis. The critical value of w for our purpose of metric engineering is

$$w < -\frac{1}{3} \tag{0.7}$$

Note that

$$w < -1 \tag{0.8}$$

is dangerous on the cosmic scale causing the "Big Rip" in our local Level I Hubble universe because the density ρ of world stuff scales as the space expansion factor $a(t)$ as

$$\rho_w \sim \frac{1}{a(t)^{3(1+w)}} \tag{0.9}$$

Therefore, the energy density explodes to infinity as the universe expands if $w < -1$ ripping the universe apart.

Starting from the SU(3) spherically symmetric quadratic interior 3D harmonic oscillator potential[xxx] for the Higgs Ocean

$$V_{zpf} = -\frac{c^2}{2}\Lambda_{zpf}r^2 \qquad (0.10)$$

The universal force per unit test mass is

$$\frac{\mathbf{F}_{zpf}}{m} = -\nabla V_{zpf} = +c^2\Lambda_{zpf}\mathbf{r} \qquad (0.11)$$

$\Lambda_{zpf} > 0$ generates a repulsive exotic vacuum force $\mathbf{F}_{(\text{Dark Energy})}$

$\Lambda_{zpf} < 0$ generates an attractive exotic vacuum force $F_{(\text{Dark Matter})}$

"What had especially caught Einstein's attention was that if Λ were ... a positive number, then the new term corresponded to a universal repulsion that increased in proportion to the distance." p. 70

"But suppose that the contribution of fermions outweighed that of bosons: then the net vacuum energy would be a negative number ... changing the sign of Λ switches the repulsive effects ... to a universal attraction: not the usual gravitational attractive force but a force that increases with distance." p. 83

Eric Davis continued: *"I'm doing a study from supersymmetric theory on the negative energy content of fermionic ZPE. The contribution from this is alleged by cosmologists to cancel out the EM ZPE above some cut-off frequency, thus yielding the tiny cosmological constant."*

No, this does not work. The cut-off frequency is still way too high. Lenny Susskind explicitly talks about this. Indeed if you read my messages carefully you will see I quote Lenny on precisely this point. Lenny wrote on this: *"Supersymmetric theories have no vacuum energy because the fermions and bosons exactly cancel. But ... Fermi-Bose symmetry is not a feature of the real world. There is no superpartner of the electron or of any other elementary particle. The vacuum energies of fermions and bosons do not cancel, and the bottom line is that our best theory of elementary particles predicts vacuum energy whose gravitational effects would be vastly too large. We don't know what to make of it. ... Finding the reason has been regarded as the biggest, most difficult problem of modern physics. No other phenomenon has puzzled physicists for as long as this one. Every attempt, be it quantum field theory or in String Theory, has failed. It truly is the mother of all physics problems."* p. 77-78, Lenny Susskind, "Cosmic Landscape."

Eric Davis *"However, no one has properly modeled the weak and strong force ZPEs, and their contributions to the EM ZPE need to be considered before we can properly address fermionic ZPE cancellation and the reason why the cosmological constant is so tiny."*

The dynamics is still supersymmetric, but the vacuum is not. It's simply "More is different" P.W. Anderson again - this is a different variation on what I previously proposed. Here the vacuum order parameter is simply proportional to the difference in the fermion and boson field species, which may vary from place to place and on different scales inside the same Level 1 Hubble universe. This would be consistent with Cosmic Landscape, i.e. tunneling from parallel "branes" into our local universe. Dominance of the fermions is attractive virtual dark matter no real particles to be detected. Dominance of the bosons is repulsive virtual dark energy. 96% of all the stuff of the world is, therefore, exotic vacuum and only ~ 4% is on-mass-shell particles whizzing through space that causes counters to click.

Experimental Concepts for Generating Negative Energy in the Laboratory
E. W. Davis and H. E. Puthoff
Inst. for Advanced Studies at Austin, 4030 W. Braker Ln., Ste. 300, Austin, TX 78759, USA 512-342-2187,
ewdavis@earthtech.org

Presented at STAIF 2006 Exotic Propulsion Section

"Abstract. Implementation of faster-than-light (FTL) interstellar travel via traversable wormholes, warp drives, or other spacetime modification schemes generally requires the engineering of spacetime into very specialized local geometries. The analysis of these via Einstein's General Theory of Relativity (GTR) field equations plus the resultant equations of state demonstrate that such geometries require the use of "exotic" matter in order to induce the requisite FTL spacetime modification. Exotic matter is generally defined by GTR physics to be matter that possesses (renormalized) negative energy density, and this is a very misunderstood and misapplied term by the non-GTR community. We clear up this misconception by defining what negative energy is, where it can be found in nature, and we also review the experimental concepts that have been proposed to generate negative energy in the laboratory."

Exotic matter is much more general then negative energy density. The pressure is important. More generally the w factor, i.e., the ratio of pressure to energy density in different directions of space.

"Exotic matter is generally defined by GTR physics to be matter that possesses (renormalized) negative energy density (sometimes negative stress-tension = positive outward pressure, a.k.a. gravitational repulsion or antigravity)"

This is suspect. Negative pressure creates repulsion. Positive pressure creates attraction.

"and this is a very misunderstood and misapplied term by the non-GTR community. We clear up this misconception by defining what negative energy is, where it can be found in nature, and we also review the experimental concepts that have been proposed to generate negative energy in the laboratory.

What is exotic about the matter (or mass-energy) that must be used to generate FTL spacetimes is that it must have negative energy density (Ford and Roman, 2003). The energy density is "negative" in the sense that the configuration of mass-energy we must deploy to generate and thread a traversable wormhole throat or a warp drive bubble must have an energy density (UE = Uc², U = mass density) that is less than or equal to its stress (W), which is written as: UE dW (Morris and Thorne, 1988; Visser, 1995)."

In the isotropic case at least, the weak field limit of GR is

$$\nabla^2 V \sim G\rho(1+3w)$$

If $\rho < 0$ & $w > -1/3$, then there is repulsion. Suppose, for example, $w = -1/4$, therefore $1 + 3w \to 1 - 3/4 = 1/4$ there is repulsion because the RHS is negative. On the other hand if $w < -1/3$ there is attraction for negative energy density. For example let $w = -2/3$, therefore $1 + 3w \to 1 - 6/3 = -1$, hence a sign change on the RHS.

Inhomogeneous Physical Vacuums

Let's look at the Newtonian limit of GR once again. Getting those damn -1's right again, if I knew those infernal -1's would cause so much trouble, I never would have quantum jumped into this Pandora's Box! ☺[xxxi] For constant uniform Λ_{zpf}

$$\mathbf{a}_{zpf} = \frac{\mathbf{F}_{zpf}}{m} = -\nabla V_{zpf} = +c^2 \Lambda_{zpf} \mathbf{r}$$

Is the poor man's "quark confinement." Start with the spherically symmetric static canonical form

$$V_{zpf} = -\frac{c^2}{2} \Lambda_{zpf} r^2 \tag{0.12}$$

But now, the torsion field extension of 1915 GR with the larger Kibble connection field from locally gauging the entire 10-parameter Poincare group allows Einstein's cosmological constant Λ to mutate into a local scalar variable field, which in this special example is a function of the radial distance r. Therefore the universal "acceleration field" (e.g, Paul Hill's book "Unconventional Flying Objects") is

$$\mathbf{a}_{zpf} = -\nabla V_{zpf} = +c^2 \left[\Lambda_{zpf} \mathbf{r} + \frac{r^2}{2} \nabla \Lambda_{zpf} \right] \tag{0.13}$$

The Newtonian weak curvature spherically symmetric static limit of this curvature-torsion generalization of Einstein's 1915 theory of gravity that only had curvature is the modified Poisson equation

$$
\begin{aligned}
\nabla^2 V_{zpf} &= -c^2 \nabla \cdot \left[\Lambda_{zpf} \mathbf{r} + \frac{r^2}{2} \nabla \Lambda_{zpf} \right] \\
&= -c^2 \left[\Lambda_{zpf} \nabla \cdot \mathbf{r} + \left(\nabla \Lambda_{zpf} \right) \cdot \mathbf{r} + \frac{r^2}{2} \nabla^2 \Lambda_{zpf} + \left(\nabla \frac{r^2}{2} \cdot \nabla \Lambda_{zpf} \right) \right] \\
&= -c^2 \left[\Lambda_{zpf} + 2 \left(\nabla \Lambda_{zpf} \right) \cdot \mathbf{r} + \frac{r^2}{2} \nabla^2 \Lambda_{zpf} \right] \\
&= -c^2 \left[\Lambda_{zpf} + 2r \left(\frac{\partial \Lambda_{zpf}}{\partial r} \right) + \frac{r^2}{2} \frac{\partial^2 \Lambda_{zpf}}{\partial r^2} \right]
\end{aligned}
\tag{0.14}
$$

Note the two new torsion field source terms on the RHS of the above equation.

NASA Pioneer Anomaly

50

For example, suppose

$$\Lambda_{zpf} = -\frac{2H(t)}{cr}$$

$$H(t) = \frac{1}{a(t)}\frac{da(t)}{dt}$$

$$V_{zpf} = -\frac{c^2}{2}\Lambda_{zpf}r^2 = +cH(t)r$$

$$\mathbf{a}_{zpf} = -\nabla V_{zpf} = -cH(t)\mathbf{e}_r \sim -10^{-7}\frac{cm}{\sec^2}\mathbf{e}_r$$

(0.15)

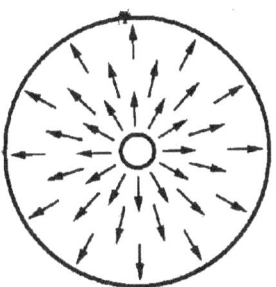

Fig. 1. Magnetization pointing outwards in the space between two spherical enclosing surfaces. This is known as a hedgehog. David Thouless Topological QM book

With a point defect in the physical vacuum order parameter at the center of the Sun that has second homotopy wrapping number 1 for a hedgehog defect starting somewhere beyond the orbit of Jupiter. We do not know how far it extends. This is a dark matter exotic vacuum hollow halo of positive quantum pressure with negative zero point energy density surrounding the Sun giving a small anomalous relative *motional* blue shift in this case because the probes are not going as fast as they would be if there were no dark matter out beyond Jupiter's orbit. On the other hand, there will also be a universal gravity red shift in this case where $\Lambda_{zpf} < 0$ superimposed upon the motional Doppler shift

$$\left(\frac{\Delta \nu}{\nu}\right)_{zpf} \sim \frac{V_{zpf}}{c^2} \sim -\frac{1}{2}\Lambda_{zpf}r^2 \sim -\frac{1}{2c}H(t)r \sim -\frac{v_{Hubble\_recession}}{2c}$$

(0.16)

This is the same order of magnitude of the Doppler shift but in opposite direction with a relative factor of 1/2 so that the relative motional blue shift effect from the additional attraction is twice as large as the gravitational red shift effect. This seems to be consistent with the observation of the net anomalous drag on the spacecraft back to the Sun giving a net relative tiny blue shift. However, these are only crude back-of-the-envelope estimates I am tentatively suggesting here.

Dark Matter Galactic Halo

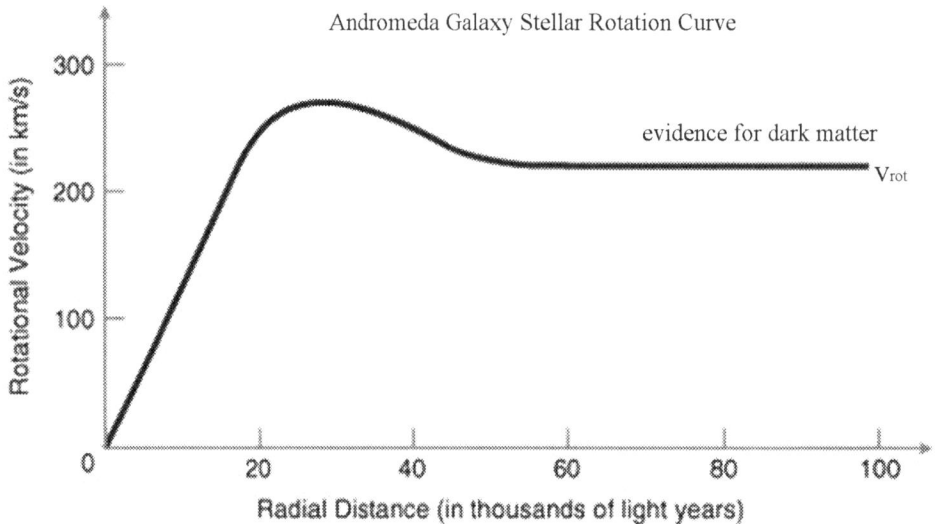

Similarly for the Galactic Halo in the "flat" part of the stellar rotation curve, let the constant circulating stellar speed be v_{star}

$$\Lambda_{zpf} : -\left(\frac{v_{star}}{cr}\right)^2 \log\left(\frac{r}{a}\right)$$

$$V_{zpf} = -c^2 \Lambda_{zpf} r^2 \rightarrow v_{star}^2 \log\left(\frac{r}{a}\right) \qquad (0.17)$$

$$\mathbf{a}_{zpf} = -\frac{\partial V_{zpf}}{\partial r} = -\frac{v_{star}^2}{r}$$

This inhomogeneous negative zero point energy distribution of positive pressure mimics w = 0 Cold Dark Matter (CDM), but it requires an additional long-range geometrodynamic field from locally gauging the full 15 parameter Conformal Group and hiding the six torsion fields from the conformal group's Lorentz subgroup. The torsion quanta get large Meissner effect masses $M_i, i = 1, 2, ... 6$ that correspond to small compactification scales

$$R_i = \frac{h}{M_i c} \qquad (0.18)$$

leading presumably to Lenny Susskind's 10^{500} vacua of pocket universes in the Cosmic Landscape of the Megaverse.

52

Extracting the Dark Energy from Space

The Bianchi identities severely restrict the local variation of Λ because the exotic vacuum field equation from Einstein's 1915 theory

$$G_{\mu\nu} + \Lambda g_{\mu\nu} = 0 \qquad (0.19)$$

Using the zero torsion Levi-Civita connection fields from the local gauging of the globally flat 4-parameter translation group T4 generated by *total* energy and linear momentum: the covariant partial derivative is $D^\nu = \partial^\nu + \Gamma^\nu \ldots$. For example,

$$D^\nu \Lambda = \partial^\nu \Lambda = g^{\nu\lambda} \frac{\partial \Lambda}{\partial x^\lambda} \equiv g^{\nu\lambda} \partial_\lambda \Lambda$$
$$g^{\nu\lambda} g_{\mu\lambda} = \delta^\nu_\mu = 1(\mu = \nu) \, or = 0(\mu \neq \nu) \qquad (0.20)$$
$$D^\nu A^\mu = \partial^\nu A^\mu + \Gamma^{\nu\mu}_\lambda A^\lambda$$
$$D^\nu A_\mu = \partial^\nu A_\mu - \Gamma^{\nu\lambda}_\mu A_\lambda$$

The equivalence principle (EEP) implies that that the universal translational inertial g-force per test mass $a^i_{inertial}, i = 1, 2, 3$

$$\Gamma^{\cdots}_{\cdots}(P)_{LIF} = 0$$
$$\Gamma^{\cdots}_{\cdots}(P)_{LNIF} \neq 0 \qquad (0.21)$$
$$a_{inertial}{}^i = c^2 \Gamma^i_{00}$$

Metricity means

$$D^\nu g_{\mu\nu} = 0 \qquad (0.22)$$

The vanishing of the independent torsion field plus metricity give the Einstein Bianchi identities that imply the vanishing covariant divergence of the Einstein tensor $G_{\mu\nu} = R_{\mu\nu} - (R/2)g_{\mu\nu}$.

$$D^\nu G_{\mu\nu} = 0 \qquad (0.23)$$

Therefore in vacuum by the product rule of differential calculus

$$g^\nu_\mu \frac{\partial \Lambda}{\partial x^\nu} = 0 \qquad (0.24)$$

This is locally a homogeneous set of 4 quasi-linear equations in 4 unknowns. However, $\det g_\mu^\nu \neq 0$, therefore

$$\frac{\partial \Lambda}{\partial x^\nu} = 0 \tag{0.25}$$

However, if there is an independent torsion field beyond Einstein's 1915 approximation to unified field theory, then Λ can be a locally variable scalar field. The dynamically independent torsion field comes from locally gauging the 6-parameter Lorentz group of space-time rotations in 4 dimensions. Furthermore, even in Einstein's theory we can have

$$g_\mu^\nu \frac{\partial \Lambda}{\partial x^\nu} = D^\nu T_{\mu\nu}(matter) \neq 0 \tag{0.26}$$

This is the two-way flow between zero point exotic vacuum currents and matter-electromagnetic field currents – a first step toward independence from oil. Locally gauging the entire 15-parameter conformal group of Roger Penrose's twistor theory gives even more physics alternatives yet to be explored.

Message from NASA AMES Scientist Creon Levit

From: Creon Levit
Sent: Sunday, April 03, 2005 8:39 PM
Subject: Re: What does Sarfatti say about this?

"I think Jack is really on to something now. He can derive GR, and hence all of classical gravitation, from a simple quantum model of the vacuum as a two-component (Ginzburg-Landau - "GL") superfluid of virtual fermion-antifermion pairs. His new theory solves the cosmological constant problem: It predicts Lambda Λ fluctuating slightly around zero (as is observed) instead predicting lambda of around 10^{66} cm^{-2} as does naive QED and superstring theory. It solves the dark matter problem: dark matter is where the "normal fluid" component of the GL fluid dominates the mixture. It solves the dark energy problem: dark energy is the converse situation - where the superfluid component dominates. It solves the inflation /arrow of time problem: the energy and the negentropy for inflation are released when the initial unstable primordial "false-vacuum" state undergoes a phase transition to the present GL state consisting of a vacuum condensate of virtual e^{+} e^{-} pairs. His new theory shows how space-time (3+1 dimensions) emerges from the large-N dimensional configuration space of quantum mechanics: through the emergence of an order parameter, in this case the giant quantum wavefunction of the cosmic GL superfluid wavefunction whose long-range order (ODLRO) correlates different positions in space and time. Inspired by the work of Hagen Kleinert, the gradient of this order parameter, and its gradient's gradient (tensor) etc., when expressed using the formalism of moving Cartan frames or Vierbeins, reproduces Einstein's equation. Gravity emerges from QM, without needing quantization itself. The allowable broken symmetries of this order parameter can explain effects like the Pioneer anomaly, the dark massive galactic halos, and the micro-Kerr-black hole model of elementary particles. When coupled with the approaches of Ray Chiao and perhaps Chapline and Laughlin it may allow the control of gravity and circumvention of the space-time stiffness barrier."

Robert Laughlin's Different Universe

"We physicists are fully aware of our own sententious tendencies and go to great lengths to keep them under control." p.x

Robert Laughlin and Lenny Susskind are both physics professors at Stanford University and they are at odds with each other on some key fundamental issues about the nature of information loss down black holes. Susskind won a recent victory at GR 17 in Dublin (July 2004) where Stephen Hawking rolled over and admitted defeat in his twenty-year debate with Susskind. On the other hand, Laughlin won the Nobel Prize in 1998 for his work on the fractional quantum Hall effect in low temperature thin films where the electrons attach to quantized magnetic vortices to form "anyons" with fractional charges like the quarks, and a weird kind of quantum statistics like Schrodinger's Cat both dead and alive. These anyons are neither bosons nor fermions and can morph into one or the other as control parameters like applied magnetic field are varied. One gets incredibly accurate and precise measurements of the ratio of the square of the electron's charge to Planck's quantum of action over a wide range of variable magnetic field for a wide range of variability of the properties of the samples. This really is "more is different" as the measurement gets less accurate and precise when the sample size is reduced. The makes entirely new phases of matter including the possibility of high temperature superconductors – really high at maybe a thousand degrees Kelvin or higher. This has not been achieved in the lab although some think that the alleged alien flying saucers use this nanotechnology. The intellectual conflict between Susskind and Laughlin must make life interesting for the students like the creative tension between Feynman and Gell-Mann at Cal Tech. They had the same secretary with offices on opposite sides of her office if I remember correctly from 1968?

What are some of Laughlin's key ideas? He comes from the tradition of P.W. Anderson's "More is different" as a general theory of emergence of levels of physical reality beyond naïve mechanical reductionism in which the higher level is largely independent of the lower but not entirely of course. This is already in Bohm's quantum potential landscape from which classical fields and particles get their marching orders. The quantum potential is form-dependent beyond the classical forces. The Bohm quantum potential landscape is Wheeler's "BIT" or QUBIT to be more precise that is physical and organizational yet non-material. It is indeed the "élan vital." The Bohm theory stops short of *adaptive conscious spontaneous self-organization* because it is the test particle approximation in which the classical fields and particles do not directly modify the landscape on which they roll. Indeed the same can be said for Lenny Susskind's cosmic landscape theory. This two-way relation tolls the death knell for unitarity and signal locality. In fact, like in Carlo Suares's "two-way" equation of Qabala shown to me by him in Paris in 1973, there is both a bottom $\rightarrow$ up and a top $\rightarrow$ down flow of influence in the Great Chain of Being and Becoming from possibly "spin foams" to curved space-time to the creation of quarks, leptons and gauge force bosons in the reheating of the big bang to the creation of galaxies from the stretched-out WMAP Gaussian scale-independent quantum gravity foam fluctuations, light atoms, molecules, stars, heavy atoms spewed out in super nova explosions, formation of planets, on to life and beyond to what to God perhaps? Don't forget the Angels and Aliens? Or is the Mind of God always about in the quad dabbling here and there in the affairs of humankind? Does God make mistakes? Getting back to Laughlin's ideas, here are some key quotations from his book:

"The laws of nature that we care about ... emerge through collective self-organization and really do not require knowledge of their component parts to be comprehended and exploited." p. xi

"I was having the same conversation with colleagues about Brian Greene's 'The Elegant Universe' ... but the subject always seemed to drift to ... the pointlessness of making models of the world that were beautiful but predicted no experiments." p. xii

Physics as Art for Art's sake - is it one more sign of the Decline and Fall of Western Civilization? Pure mathematicians encroach on theoretical physics?

"Moreover, it was an ideological dispute: it had nothing to do with what was true and everything to do with what 'true' was."

"Organization can acquire a meaning and life of it's own and begin to transcend the parts from which it is made." p. xiv

"The distinction between fundamental laws and the laws descending from them is a myth, as is the idea of mastery of the universe through mathematics alone."

"Physical law cannot generally be anticipated by pure thought, but must be discovered experimentally, because control of nature is achieved only when nature allows this through a principle of organization." p. xv

Indeed, Lenny Susskind laments the lack of such a principle of organization for M-Theory formerly called string theory but extended now to a brane new world. Einstein's equivalence principle seems to me to be the missing organizing idea when it is formulated as the universal local gauging of all space-time symmetries of the physical action. Einstein's original use of the equivalence principle in 21st Century terms was simply the local gauging of the 4-parameter translation group generated by the total energy and linear momentum. However there are at least 11 other generators in what is called the Conformal Group." Lenny seems recently to have found the Lost Chord in his cosmic landscape with eternal inflation populating its every nook and cranny giving us 10^{500} parallel pocket universes, which is quite enough to explain the fine-tuning of the cosmological constant out to the 120th decimal place in terms of the Weak Anthropic Principle. What of God, the Intelligent Designer? Lenny quotes Lagrange to Napoleon "I have no need of that hypothesis."

Nonlocally Coherent Locally Random Zero Point Energy in Superfluids

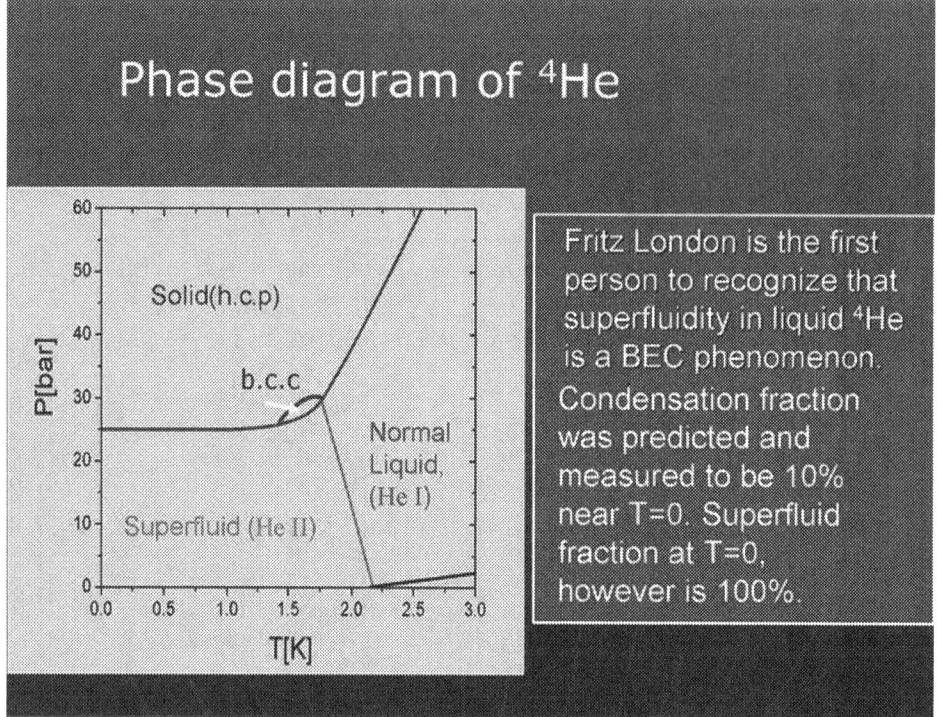

Lecture by Moses Chan[xxxii]

Note added to this revised edition of December 2005. Through e-mail with Dr. Robert Becker, there has been a new development. The microscopic many-body quantum theory shows that the ODLRO Bose-Einstein condensate is only a few percent of the total density in models of integer spin bosons with forces between them in contrast to an ideal gas in which there is only the quantum nonlocality of permutation symmetry operating. Note that quantum nonlocality is evident even in the simple double slit experiment with a single photo passing the slits at a time. With both slits open a photon never lands in a region of destructive interference on the distant screen. With only one slit open it does land there. The photon leaves a local trace on the screen as if it is a tiny particle. Yet the tiny photon passing one open slit knows instantly that the other slit is either open or closed. That is a good example of quantum nonlocality acting faster-than-light over the interval between the slits when the photon allegedly passes only through one of them. Returning to the problem that the condensate density is only a small fraction of the total density, the problem is what is the nature of the remaining density that is most of the stuff at near absolute zero temperature? It's the zero point quantum jiggle or jitter of the particles. Robert Becker has suggested that this zero point jiggle while locally completely uncontrollably random is, nevertheless, nonlocally phase-locked or coherent. This is precisely what happens in the Einstein-Podolsky-Rosen pair state in Alain Aspect's famous Paris experiment in 1982 or so. Each photon in the pair is locally randomly unpolarized, but there is a definite nonlocal polarization correlation. Similarly if you put interferometers at each end of a photon pair experiment there will be no local fringe patterns, but you will see nice sharp nonlocal fringe patterns when you correlate the data from both ends of the system. You *cannot* use this as a faster-than-light signal device without violating quantum theory. That's the "no-cloning" theorem of "signal locality" in nonlocal micro-quantum theory. Macro-quantum theory, however, may be a very different story. In any case, the puzzling fact was that the G-L phenomenological semi-empirical theory says that the total density at near absolute zero temperature is 100% coherent superfluid. This can only be so if the locally random zero point motion is also nonlocally coherent in the sense of the Einstein-Rosen-Podolsky phase-locking that we see in the above experiment using two interferometers at each end. That is, the total GL superfluid density is the sum of the condensate density plus the nonlocally coherent zero point density.

$$\rho_s = \rho_0 + \rho_{zpf} \tag{0.27}$$

This same equation inside the vacuum for virtual stuff comes up in my theory of dark energy and dark matter as negative and positive zero point quantum pressure phases of the physical vacuum respectively. Note that when the temperature is above absolute zero there is a third fluid component called "normal fluid" of elementary excitations and collective modes outside the vacuum. However, in this book we are mostly interested only in the virtual processes inside the vacuum.

Making Star Trek Real

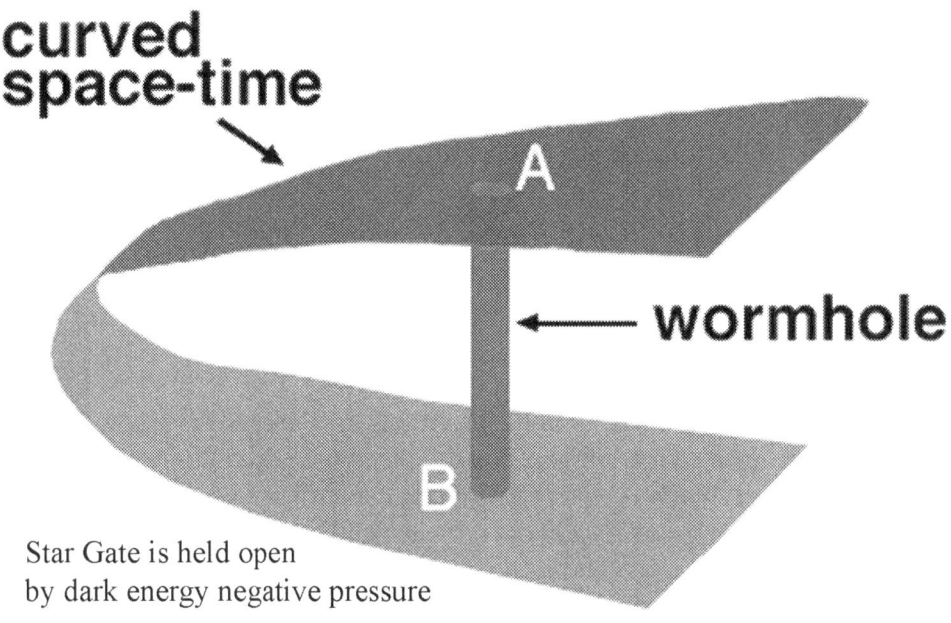

curved space-time

A

← wormhole

B

Star Gate is held open
by dark energy negative pressure

Precision cosmology data from 1999 - 2004 on Type 1a supernovae, gravity lensing and the NASA WMAP space probe have shown that our universe is spatially flat and is accelerating in its expansion rate from an anti-gravity "dark energy" field that is 73% of the large-scale stuff of the world. Recent experiments (CDMSII) show that very sensitive dark matter detectors do not click as I predicted they would not. Ordinary matter made from electrons, neutrinos, nucleons and photons in gas clouds, stars, planets, black holes et-al are at most only 4% of all the stuff in the universe. I have shown that both dark energy and dark matter accounting for 96% of the stuff of the universe are different phases of exotic vacuum with negative and positive quantum pressures respectively in accord with standard results of Einstein's theory of gravity and quantum theory. Elementary particles like the electron have a strongly gravitating zero point energy inner core of exotic vacuum that prevents the self-electric charge from exploding just as the invisible galactic halo, also a sphere of zero point energy of positive quantum pressure, prevents the stars in our galaxy from escaping willy-nilly. As Above, So Below. I have also shown how, in principle, using the Bohm-Aharonov-Josephson effect with rotating superconductors (E. Podkletnov & Ning Li), we can manipulate the relative amounts of anti-gravitating dark energy and gravitating dark matter to make stable traversable wormhole star gates for effective global super-fast space travel and even time travel, while hardly moving at all locally, within our universe and the parallel universes next door in the hyperspace of super cosmos. The dark energy is what Herman Bondi called "negative matter" and can, therefore, be used for "acceleration field" (Paul Hill) weightless warp drive propulsion in Alcubierre's sense with zero g-force and small tidal curvature stretch-squeeze even in high speed hairpin turns. This implements George Trimble's 1956 concept of the "G-Engine" for Unconventional Flying Objects discussed in Nick Cook's "The Hunt for Zero Point" from Jane's Defence Weekly and William

Austin's article "To The Stars" in Aviation Week and Space Technology March 1, 2004. The military implications for air and space defense and WMD of these new physics concepts are obvious. There have been recent claims as this book goes to press that the wormholes are unstable. The UFO evidence suggests that is false. Other physicists are independently thinking along similar lines to me:

The Rolling Stones

"The Landscape is a space of possibilities ... It has geography and topography with hills, valleys, flat plains, deep trenches, mountains and mountain passes. But unlike an ordinary landscape, it isn't three-dimensional." P. 90 "Cosmic Landscape"

In contrast to Lenny's micro-quantum substratum vacuum landscape, the macro-quantum landscape is 3D. This is why space-time physics is local and why we do not find giant Schrodinger Cats in a coherent superposition of dead and alive. Note that Bohm's hidden variable theory, like Murray Gell-Mann's complexity theory, uses a landscape of many dimensions as well. All BIT pilot waves, excited states as well as the ground state are landscapes on which the hidden variable IT "system point" is a rolling stone that gathers no moss. Lenny's many-dimensional vacuum landscape is BIT telling IT how to move without any direct back-reaction of IT on BIT. This implies "no cloning" i.e., "signal locality" without any possibility of using nonlocal quantum entanglement to send messages faster than light or backward through time in telepathic controlled spooky action at a distance.

"The Landscape ... doesn't exist in space and time at all ... each of [its] points represent a possible ... vacuum ... a potential for all the things that can happen in that background. It mean a list of all the elementary particles as well as the constants of nature ... it means an environment in which the Laws of Physics take a particular form." P.91

UFOS: Is War of The Worlds Real?

National Security Agency Website on UFOs

On Feb 12, 2006, at 9:57 AM, Ted Roe of NARCAP wrote:
"Howdy, Go to the NSA website[xxxiii] and review Items 17 and 20 for a basic overview... I understand that Richard Secord was in charge of the follow-up investigation. I don't think anyone has seen that report."

On Feb 12, 2006, at 10:05 AM, Bruce Maccabee wrote:
"You asked about the CIA's explanation as referenced by Stephen Schwartz who used the name Haines as the source. There is a reason for using the name Haines: the author was the former CIA historian Gerald K. Haines. You can find out why his explanation is preposterous by reading my website." [xxxiv]

On Feb 12, 2006, at 9:35 AM, Bruce Maccabee wrote:
Message text written by Jack Sarfatti
Please explain in much more detail what you mean? "Interference"?
On Feb 10, 2006, at 9:52 PM, Bruce Maccabee wrote:
"If UFOs/ET's have the capabilities exhibited during the Iranian Jet/UFO chase case of Sept 19, 1976 (before the Ayatollahs took over), then (our) space weapons had better be pretty damn good at rejecting (ET) "interference" with their operation. Two, F-4 jets, were directed, by an Iranian Air Force general to chase a flashing light UFO over Teheran. The jets were launched ten minutes apart from an airfield about 100 miles
away. Each jet chased the object at speeds up near Mach 2, and approached to within 25
nautical miles (approx) and couldn't get any closer. The first jet, upon reaching about 25 nm suddenly lost all electronics and communications including the intercom with the "back-seater." The jet turned away and regained electronics, etc. and went back to base. The second jet was launched 10 minutes after the first. It approached the speeding object and radar showed closure until the distance shrank to about 25 nm. Object emitted a smaller bright object. The pilot decided to arm AIM9 heat seeking missile. At that "instant" he lost all electronic control and communications, not just with the ground, but also the intercom between the pilot and the "back-seater." The plane turned away and regained all electronics. That's what I meant by interference."

Ron Stahl wrote:
I read the report[xxxv] a little more carefully and saw no mention of the F-4's speed. It was carrying AIM-9's which it locked just before all electronics went out. Robert, if you were flying in '76, maybe you can throw light on what they're talking about that the object was similar in size to a "T-Bird". Was the T-33 notably smaller than an F-4?

On Feb 12, 2006, at 12:57 PM, Captain Robert M. Collins, USAF Intelligence, Retired, author of "Exempt From Disclosure" wrote: "Since the AIM -9 is heat seeking and if you get a lock for just a few seconds you fire the missile even if your radar goes out. That F4 exceeded Mach 1 & 2 the way I understand it for just brief moments, which is good or that Iranian pilot would have ripped the missile off his wing pylon. And yes, the T33 is definitely small than an F4."

On Feb 12, 2006, at 10:42 AM, Michael Salla wrote:

"Hello Bruce and all, this is a two-way process. In his original notes (published in Italy) (Colonel Philip Corso) describes an encounter at Red Rock Canyon with a Gray EBE. Corso at the time was Battalion Commander at White Sands (1957-58). He was asked by the EBE to turn off his battalion's radar for a short period. Corso asked, "What's in it for me?" "A New World if you can take it," replied the EBE. Shortly afterward, a UFO was sighted taking off from the area. This is confirmation that radar does interfere with the electronics and navigation systems of the EBE craft. On the other hand, Capt Robert Salas, Robert Jacobs and others have described how UFOs can interfere with the electronics of missile systems both prior to and during launch, as well as the instruments of jets as you describe in the Iran incident. Corso in interviews described how building some protection for the launch and guidance systems of weapons from 'interference' has been going on for some time. He described such a conversation with Edward Teller where the problem of defending electronic systems from EMP was discussed. Some breakthroughs may have occurred with protecting conventional weapons systems such as ballistic missiles and jet launch systems, but I'm sure this would be highly classified. However, from the public information available, both the delivery and navigation components of missile systems can be 'influenced' by the EBEs. However, when it comes to Directed Energy Weapon systems, as described in Douglas Beason's book, "E-Bomb", we have an entirely new generation of weapons being developed and deployed. With directed energy weapon systems, there is no guidance system as in a plane or missile that can be interfered with after launch. Once a laser, particle beam weapon or microwave is set off, it travels to the target at close to the speed of light and the EBE craft would have to take evasive action. Hence, the likelihood of 'interference' only applies at the ground level or satellite where the energy weapon is launched in the first place and I'm sure these have been 'protected' as far as possible against the kind of interference that you describe."

On Feb 12, 2006, at 10:55 AM, Captain Robert M. Collins, wrote: "Won't do any good, they already had all those protections for missile silos going way back, yet UFOs could shut down a single missile or an entire missile field. Our Military is living on pipe dreams."

With metric engineering using controlled dark energy via phase-locking of probably a 2D anyon "shield" to the local vacuum coherent phases you can change the geodesic paths of the directed energy beam. But Salla is probably correct that that the alien enemy has some kind of pulsed EMP beam weapon like the "photon torpedo" in Star Trek.

Colonel Philip J. Corso finds an Alien Time Machine?

On Feb 12, 2006, at 1:37 PM, William Birnes wrote:

"Hi, Jack,

Thanks for copying me on this. There's a whole back-story to this incident. When Corso was at the Army's Red Canyon facility, one of the first things he was told was that he would receive orders to shut down his radars at various times and that he had to be exact about the times. But these were odd times and made no sense to him. As this continued, he off-handedly suggested to one of his radar operators that the operator might fudge a little on the time, wait a couple of seconds after the shut-down time before shutting the radar off. And while he was at the console, he might also fudge a little before turning them back on, maybe a few seconds early. What they picked up on the screen, Corso said, were hits on strange objects. Corso didn't know what they were and didn't directly question anyone about it. But he guessed that at a base command level, they knew what was in the sky and were cooperating with them for some reason. Corso suggested to me that this smacked of some sort of arrangement with whatever those objects were to conceal them from radar operators and other personnel not in the loop.

The actual encounter with the EBE, Corso said, took place in a cave near where he had seen a strange object on the ground during one of his range fly-overs. When he came upon this EBE in the cave, he drew his weapon and demanded "friend or foe." The EBE communicated to him (telepathically?), "Neither," which Corso said made no sense to him at the time. But the EBE asked Corso to holster his weapon, to which Corso replied, "What's in it for me?" That's when the EBE told him, "A new world if you can take it." I believe that Corso first told this story to George Knapp.

The object on the ground, Corso said, was shimmering, but it was ice cold to the touch. Then, it seemed to wink out of existence. Years later, one of his scientific advisors at Army R&D told him it was a time machine. That stuck with him for the rest of his life.

In 1997, when we were filming the "Dateline" interview with NBC correspondent John Hockenberry in the Roswell museum and were passing the display of the craft's crash into arroyo, Corso said to Hockenberry that it was actually a time machine. Hockenberry wanted the whole story, but only the time machine comment got edited into the final piece. Corso's "Dawn Of A New Age" will soon appear in its original English version." [xxxvi]

From the Real X-Files of Dr. Bruce Maccabee

"The Director of Intelligence advises that no theory exists at the present time as to the origin of the objects and they are considered to be unexplained. Much of the publicity has been based on authorized news releases by the Air Force."
Gilbert Levy, Chief of the Counter Intelligence Division of the Office of the Inspector General of the US Air Force Office of Special Investigations, July 29, 1952[xxxvii]

"A Navy photographer, while traveling across the United States in his own car, saw a number of objects in the sky, which appeared to be flying saucers. He took approximately thirty-five feet of motion picture film of these objects. He voluntarily submitted the film to Air Intelligence who had it studied by the Air Technical Intelligence Center. Experts at the Air Technical Intelligence Center have advised that, after careful study, there were as many as twelve to sixteen flying objects recorded on this film; that the possibility of weather balloons, clouds or other explainable objects has been completely ruled out; and that they are at a complete loss to explain this most recent creditable sighting. The Air Technical Intelligence Center experts pointed out that they could not be optical illusions inasmuch as optical illusions could not be recorded on film." FBI Memorandum, Oct. 27, 1952

"The Air Force is mostly interested in the 'saucer' problem because of its psychological warfare implications... On the other hand, several 'saucer' societies in the United States have been investigated. Key members of some of these societies, which have been instrumental in keeping the 'flying saucer' craze before the public, have been exposed as being of doubtful loyalty. Furthermore the societies, in some cases, are financed by an unknown source." CIA Memorandum, August 22, 1952

"OSI (Office of Scientific Intelligence within the CIA) entered into its inquiry fully aware that it was coming into a field already charged with partisanship, one in which objectivity had been overridden by numerous sensational writers, and one in which there are pressures for extravagant explanation as well as for oversimplification. The OSI Team consulted with a representative of the Air Force Special Studies Group; discussed the problem with those in charge of the Air Force Project at Wright Field; reviewed a considerable volume of intelligence reports; checked the Soviet press and broadcast indices; and conferred with three OSI consultants, all leaders in their scientific fields, who were chosen because of their broad knowledge of the technical areas concerned. OSI found that the ATIC study is probably valid if the purpose is limited to a case-by-case explanation. However, the study makes no attempt to solve the more fundamental aspect of the problem, which is to determine definitely the nature of the various phenomena which are causing these sightings, or to discover means by which these causes and their visual or electronic effects may be immediately identified. Our consultant panel stated that these solutions would probably be found on the margins or just beyond the frontiers of our present knowledge in the fields of atmospheric, ionospheric, and extraterrestrial phenomena, with the added possibility that our present dispersal of nuclear waste products might also be a factor."

Dr. H. Marshall Chadwell, the Assistant Director for Scientific Intelligence, CIA in September 1952.

"Recent reports reaching the CIA indicated that further action was desirable and another briefing by the cognizant A-2 (air intelligence) and ATIC personnel was held on 25 November. At this time, the reports of incidents convince us that there is something going on that must have immediate attention. The details of some of these incidents have been discussed by AD/SI (Associate Director of the Scientific Intelligence Division) with DDCI (Deputy Director of the CIA). Sightings of unexplained objects at great altitudes and traveling at high speeds in the vicinity of major U.S. defense installations are of such nature that they are not attributable to natural phenomena or known types of aerial vehicles."

Dr. H. Marshall Chadwell, the Assistant Director for Scientific Intelligence, CIA in December 1952.

Dr. Bruce Maccabee, a physicist who worked in the US Navy, has contributed a detailed original article on the evidence for nuts and bolts flying saucers with warp drive and possible time travel capability. He clearly shows the early interest of the American Government in the phenomenon whose basic physics I explain in this book whilst also giving space to the loyal opposition of those who deny the reality of the UFO phenomenon as an advanced non-human super-technology that can lay the best laid plans of mice and men awry potentially rendering our entire military force "impotent and obsolete."

"Bruce spent his early years in Rutland, VT. After high school he studied physics at Worcester Polytechnic Institute in Worcester, Mass (B.S. in physics) and then at The American University, Washington, DC (M.S. and Ph. D. in physics). In 1972 commenced his long career at the Naval Surface Warfare Center, presently headquartered at Dahlgren, Virginia. He has worked on optical data processing, generation of underwater sound with lasers and various aspects of the Strategic Defense Initiative (SDI) and Ballistic Missile Defense (BMD) using high power lasers."

An American Demonology

"Captain Edward Ruppelt was head of the US Air Force UFO investigations in the early 1950s and author of "The Report on Unidentified Flying Objects, the first book to look into what President Eisenhower termed 'the military industrial complex' and how it dealt with this strange, new phenomenon. An American Demonology is not a book aimed at research specialists. Easy to read, it discusses the work of Captain Ruppelt, who tried desperately to convince the US Air Force that UFOs were real. It also takes a broad look at the politics of American technology and science within the military culture of the era, sightings of UFOs over Washington in the summer of 1952, Project Grudge and the burgeoning media interest in flying saucers."

"An American Demonology" Colin Bennett & Nick Pope May 2005

On Dec 17, 2005, at 4:10 PM, Ed Komarek wrote:
"I ran into this article in Nexus Magazine about a UFO landing on the Capitol Roof with pictures taken by a high level employee of ABC News. He claims to have worked at the White House and on Air Force One. These pictures may have some useful information as to the propulsion systems of craft as explained in the article."[xxxviii]

The pictures are not very good.

"Jack run down the article to the really good pictures starting on page six."

OK, I saw them.

"Good this maybe something you can use."

The color patterns are curious. But are those close to actual colors or is it Photoshop?

Blue shifts mean repulsive ZPF induced gravity from that patch of the craft, red shifts mean attractive ZPF induced gravity from that part of the craft. The craft should move in the direction from blue to red - if my theory is correct. That's opposite to motional Doppler shift of course.

The size of the shift is

$$\frac{\Delta\omega}{\omega} \sim \frac{V_{zpf}}{c^2}$$
$$\nabla^2\left(\frac{\Delta\omega}{\omega}\right) \sim \nabla^2\left(\frac{V_{zpf}}{c^2}\right) \sim \Lambda_{zpf}$$

(0.28)

For flight in air on Earth we need

$$\Lambda_{ZPF} \sim 1/(10^{13}\ cm)^2 = 1/r_{curvature}^2$$

$$GM_{Earth}/c^2(radius\ of\ Earth)^3 \sim 1/\ r_{curvature}^2$$

68

Note if it's ordinary G inside the thin films embedded in the fuselage (with array of tiny superconducting coils)

Then the ZPF energy density inside the thin films is $\sim (10^{49}$ ergs/cm$)(10^{-26})$ cm^{-2} $\sim 10^{23}$ ergs/cm^3. The trick is to tune the effective G* via some hyperspace effect. See Lisa Randall's "Warped Passages" for an intro to how maybe something like that can work.

The effective Planck area L_p*^2 scales as G*.

It also scales as $1/c^3$ and Hal Puthoff wants to change c in vacuum I think.

$$L_p*^2 = hG*/c^3$$

Therefore, if you make c smaller inside the thin film then you increase L_p*^2 which is what you want to do. That makes space-time less stiff, which means it's much easier to warp. of course c slows down in an atomic gas Bose-Einstein condensate - so that's a clue since the vacuum also has a condensate along with locally random but nonlocally phase coherent ZPF.

The Secret of the Flying Saucer?

"The spin-statistics theorem only holds for space-times of dimension 4 and up. ... Take, for example, a thin 2-dimensional layer of stuff: this can act like a little 3-dimensional space-time. Similarly, filaments can act like 2-dimensional universes. These days condensed matter theorists delight in the odd processes that occur in these contexts, and it was only a matter of time before someone noted that one can, at least in principle, arrange to get particles that are neither bosons or fermions. Wilczek is generally credited with taking the idea of these anyons seriously, though it had occurred to others earlier."
John Baez

"An anyon is a type of projective representation of a Lie group... These representations are called anyons. The topological reason behind the phenomenon is this: the first homotopy group of SO(2,1) ... is Z (infinite cyclic). This means that Spin(2,1) ... is not simply connected. ... This mathematical concept becomes useful in the physics of two-dimensional systems such as sheets of graphite or the quantum Hall effect. In space of three dimensions (or more), elementary particles have tightly constrained quantum numbers and, in particular, are restricted to being fermions or bosons. In two-dimensional systems, however, quasiparticles are observed whose quantum states range continuously between fermionic and bosonic, taking on any quantum value in between. Frank Wilczek coined the term "anyons" in 1982 to describe such particles." Wikepedia online.

The anyon, roughly speaking is a composite of an electron with quantized magnetic flux strings stuck to it. The anyons come in three phases: metal, superconductor and liquid. The superconducting phase, in principle, can exist up to a thousand degrees Kelvin, although this has not been achieved in the laboratory. Are we seeing this high temperature anyonic thin film superconductivity in the skins of the flying saucers?

"Surely the most dramatic result of the study of anyon statistical mechanics has been the demonstration of a new mechanism of superfluidity (and for charged anyons, superconductivity). This superfluidity is quite a robust consequence of fractional quantum statistics at appropriate values of the (filling) fraction. It occurs even in the presence of other repulsive interactions, and gives quite a large energy gap." Frank Wilczek

The larger the energy gap the higher the temperature at which the system superconducts.

"It is essentially two-dimensional and plausibly associated with exotic spin ordering."

"The spin liquid state of a Mott insulator, hypothesized by P.W. Anderson and identified by him as ... high temperature superconductors ... exhibits the ... fractional quantum Hall effect ...particles carrying a fraction of an elementary quantum number, in this case spin, attract one another by a powerful gauge force, which can lead to a new

kind of superconductivity." R. B. Laughlin, "The Relationship Between High Temperature Superconductivity and the Fractional Quantum Hall Effect", Science, 242, Oct, 1988.

On Jul 9, 2005, at 10:32 AM, Jack Sarfatti wrote:

The stuff below may be relevant. The issue is how far the formal correspondence between the two systems goes. P.W. Anderson says that once you have ODLRO the micro details don't matter that much. The issue then is do the different kinds of ODLRO fields see each other to couple? That is related to what Ray Chiao at UCB calls "impedance matching" in the context of his idea of "gravity radio" where the 3-vector gravimagnetic field g_{0i} couples to the EM vector potential A^i i.e. a $g_{0i}A^i$ interaction Hamiltonian in the NR approximation with a local macro-quantum ODLRO field Ψ of a type II superconductor with self-trapped quantized magnetic flux vortex filaments where the interaction Hamiltonian is effective. The same thing happens in laser nonlinear optics as I published in 1967 read by Ray Chiao back then. Now in the conventional Josephson junction (JJ) you have two identical superconductors with ODLRO fields Ψ_1 and Ψ_2 separated by a thin normal metal barrier. The basic JJ effect comes simply from the giant quantum wave interference $|\Psi_1 + \Psi_2|^2$ see Feynman Vol III. It does not matter much that Ψ in this case comes from real electron pairs bound together by a virtual phonon that is able to overpower the repulsion from the virtual photon in the dominant Feynman diagrams.

In my picture of the post-inflation Higgs Ocean, there is a vacuum ODLRO BEC macro-quantum "condensate" of mainly virtual electron-positron pairs bound together by a virtual photon. The only issue then is whether the Goldstone phase $\arg \Psi_{vacODLRO}$ can interfere with some ODLRO field such as, for example, Ψ_{anyon} in the fuselage of the "saucer". I guess anyons because they exhibit high-temperature macro-quantum ODLRO in thin 2D layers consistent with Colonel Philip J. Corso's reports of "thin aluminum-like foil" saucer fuselage debris, but of enormous strength from the Roswell crash in 1947. Colonel John Alexander has assured me of Colonel Corso's personal integrity and honesty. Also other things Corso said fit. Since the saucer is allegedly guided by the EEG brain waves of the AI "Gray" that the thin nano-engineered hull is also a quantum computer is a very nice feature.

On Jul 8, 2005, at 6:03 PM, Jack Sarfatti wrote:

Yes an array of tiny LCR loops embedded in the thin sheet. Something like that. Basically a lot of tiny Josephson junctions coupling the vacuum ODLRO to the high Tc anyon ODLRO rather than a thin normal barrier between 2 superconductors. The latter couples causes real pairs to jump through the barrier. Here we have a kind of phase lock between the vacuum coherence and the 2D anyon coherence. The idea for low-power metric engineering is to make Λ_{zpf} positive and negative in different parts of the thin shell fuselage using electric and magnetic near fields impressed on the control surface.

e.g. the phase difference $\sim 2\pi$ (magnetic flux through a small loop)/(quantum of magnetic flux)

"There's plenty of room at the bottom." Richard Feynman

We need to control electrostatic fields and magnetic fields inside the thin 2D anyon layer impedance matched to the vacuum coherence. We want the LCR oscillator over damped, i.e. $\omega^2 < 0$ mostly I suspect rather than real high Q oscillations. You get quick turns of the saucer without pulling g-force by quick relative phase changes over different parts of the bottom $\rightarrow$ up nano-engineered thin shell active layer "painted on" the fuselage - something like that.

Apr 6, 2005, at 1:03 PM, Art Wagner wrote:

"Jack, a while ago I suggested On you take a look at "Gravitational Anyons" by Stanley Deser ... *'antigravitating layers driven by pressure terms'* See what you see."

I knew him well Horatio at Brandeis. In any case it's obvious that 2D high T_c macro-quantum thin foil is how The Grays do it! Remember I don't think there are gravitons anymore. I don't need gravity anyons. Ordinary ones will do nicely, thank you.

OK here is what is obvious to me.

1. Assume all the data from Colonel Philip J. Corso is essentially correct - making allowances for his lack of knowledge of physics of course.

2. The active material of the saucer fuselage is very thin like aluminium foil, but strong. OK it's nano-engineered that's obvious. Why thin? Obvious also. You want a 2D system like anoyons, quantum Hall effect, high T_c You want high T_c obviously.

3. In my theory the induced quantum pressure inside the "foil" is proportional to

$$\Lambda_{zpf} \sim \left|\Psi_{Anyon}\right|\left|\Psi_{VacuumODLRO}\right|\cos \lim_{Loop \to 0} \int_{Closed-Loop} \left(\left(\frac{2e}{hc}\right)A_{\mu Anyon} - B_{\mu VacuumODLRO}\right)dx^{\mu}$$

i.e.

$\Lambda_{zpf} \sim$ |Anyon Condensate Density|$^{1/2}$|Vacuum ODLRO Density|$^{1/2}$ Cos[Loop Flux Integral]

$\Lambda_{zpf} > 0$ is negative quantum pressure making artificial warp anti-gravity blue-shift and repulsive Paul Hill "acceleration field" in the exterior of a compact distribution.

$\Lambda_{zpf} < 0$ is positive quantum pressure making artificial warp gravity red-shift and attractive Paul Hill acceleration field etc.

Well this is all you need for George Trimble's "G-Engine" 1956 and Bob Forward's "Negative Mass Propulsion" already known to Hermann Bondi & Stalin's Science Spy

Chief Y. Terletski 50 years ago, but in a useless formulation because they did not know of dark energy that would not be discovered until 1999 or so. Of course the "ET" know or they would not be here. Remember Michio Kaku is saying very similar things without my details. It's all you need to make Star Gates with time travel is my bet. In any case, this is fun! :-)

Are Indian Defense physicists using my dark energy UFO model?

I am told that the Tata Institute is working on this. On Mar 17, 2005, at 3:14 PM, Gary S. Bekkum wrote: Some one at India Daily continues to run with a series of stories on dark energy UFO propulsion:

UFO Propulsion System - Bending time and space[xxxix]
Staff Reporter
Mar. 17, 2005

"Defense Scientists and Engineers are finally starting to understand the ultimate propulsion system that can make instantaneous space travel possible. The UFOs use this propulsion system to travel great distances instantaneously … Bending space and time is the concept where you do not travel to the destination; you bring the destination close to you. This is exactly how UFOs travel from one destination to another. That is the reason why those who are waiting with their telescope in the open sly to find UFOs will never find them. Scientists and Engineers are now perplexed trying to understand how that kind of space travel is possible. The answer lies in using something known as dark energy ... Physicists and Engineers are researching the use of dark energy. The purpose is to understand the process of bending the space and time. That requires UFO flight patterns; sighting information and a mathematical algorithm to back calculate their flight positions in a more than three dimensional geometry. … Sometimes you wonder what any one can gain spending so much time and money to hide and cover up all the UFO information? Why so much ridicule around something every country in running after since 1890? The answer is the fact that any one who can master this time and space bending technology will be ahead of others by many years. There are early indications that scientists and engineers have got an early indication to the concept of bending time and space using dark energy. Interestingly, the whole concept starts with Einstein's Theory of Relativity

PSI Wars!

"Dr. Sarfatti, first I want to tell you that I am a Qabalist." (significant arching of eyebrow and wink) *"Dr. Sarfatti, it is my duty to inform you of a psychic war raging across the continents between the Soviet Union and your country and you are to be in the thick of it."* The late Dennis Bardens, BBC producer, author of biography of Winston Churchill, paranormal researcher, to me at Blue Boar Inn, Cambridge University Spring, 1974. The rest of this amazing true adventure Cold War spy story with Uri Geller, Werner Erhard and others is in my autobiography "Destiny Matrix."

Project Star Gate

3-6/50

26 DEC 63

MEMORANDUM FOR: Deputy Director for Science and Technology

SUBJECT : Clandestine Services Requirement for Research in the Life Sciences

1. Recently, reports from various sources on life science research in the Soviet Union have been called to my attention. These reports indicate a current preoccupation by an important sector of Soviet biological science with cybernetics, telepathy, hypnosis, and related subjects. Stimulated by these reports, I would like to pass on to you some thoughts on the possible significance of these activities to the Clandestine Services.

2. Prior to the creation of the Office of the DDS&T, a small group in TSD attempted to follow current research in the life sciences. Their particular interest was in the kind of scientific progress that indicated the feasibility of developing operationally useable systems. After ten years of following this subject, TSD's reading was that, with minor exceptions, the fields of hypnosis, telepathy, and general control of human behavior were not ready for operational applications.

3. I feel there is a need for a continuing search and reexamination of this somewhat esoteric (and perhaps scientifically disreputable) range of activities toward the end of useable techniques for the CS. TSD's current mission has been oriented away from this sort of basic research activity and towards close-in support of clandestine agent operations. In the light of the reports mentioned above, I am concerned that recent reported advances in these fields may indicate more potential than we believed existed. I can perceive operational applications, for the CS, in the areas of agent communication, interrogation, telemetry, and in the general area of control of human behavior, should breakthroughs in our understanding of these phenomena occur.

4. I am assuming that your research group is taking this area of activity under their cognizance and they will, through their contacts in TSD, bring to our attention any indications of practical

utility that can be made of these techniques so that TSD can proceed toward specific operational applications.

5. I am happy to note that TSD and ORD are proceeding with a series of monthly meetings to developing an understanding in depth of CS requirements in these and other fields. This note is not meant to supplant such a series of discussions as it applies to behavioral activities. I merely want to reinforce the dependence of the CS on your office for these matters.

SGFOIA3

Richard Helms
Deputy Director for Plans

Distribution:
    O & 1 - Addressee
        2 - DDP
        2 - TSD

DEFENSE INTELLIGENCE AGENCY

WASHINGTON, D.C. 20340-

S-23,025/PAX-TA

10 May 1995

MEMORANDUM FOR THE DIRECTOR OF RESEARCH AND DEVELOPMENT, CIA

SUBJECT: Declassification and Cooperation in Response to Congressionally Directed STAR GATE Program Review (U)

Reference: ORD/CIA Draft Memorandum, 21 February 1995, subject as above.

1. (C/NF) Any decision on declassification actions regarding Project STAR GATE should be made by the CIA. Executive Order 12958 "Classified National Security Information," states that in a transfer of functions, the receiving agency is considered to be the originating agency for any actions affecting program status.

2. (S/NF) DIA has no objection to a declassification of the "fact of" a DIA program to employ paranormal phenomena for intelligence purposes. If CIA decides to declassify this, or any other DIA-specific aspects of the STAR GATE project, please notify this Office two weeks in advance so that we can take appropriate steps to prepare our Public Affairs Office.

FOR THE DIRECTOR:

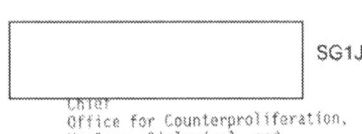

SG1J

Chief
Office for Counterproliferation,
Nuclear, Biological, and
Chemical Assessments

D
G
I

DEFENSE GROUP INC

Corporate Office: 337 Annandale Road, Falls Church

SG1B

27 February 1995

Mr. Andrew W. Marshall
Director Net Assessment
Room 3A930, Pentagon
Washington, D. C. 20301-2950

Dear Andy:

Per guidance from Lin Wells, the enclosed proposal (your copy) is to address the sensitive matter discussed with you and Lin Wells earlier, and verbally approved.

I have had the document written in response to your recent BAA whose deadline is Tuesday, 28 February 1995, with delivery to Glenna Hughes. I have included enough "codeword" phrases to identify the thrust of the work - but not the scope nor the totality.

Please call if I can clarify any issues.

Sincerely,

James P. Wade, Jr.
Chairman of the Board and
Chief Executive Officer

Enclosure
As Stated

Copy furnished:
Glenna Hughes

Approved For Release 2003/04/18 : CIA-RDP96-00789R002600030012-8
- ~ Mail ~ -
September 22, 1994 12:17pm MAIL IS -
FROM: Jay Sloan   Viewed Ntfy Private Urgent
TO:

SUBJECT: SG Briefing for Andy Marshall
COPY:

John T. Berbrich

Lin Wells has asked us to set up a meeting with himself and Andy Marshall for the purpose of briefing Andy on Stargate.

I'm not certain of the "protocol"; i.e., should the meeting be in Lin's office or in Andy's office? Plse coord with PSX

Briefing requirement should be tasked to   as senior POC's. Briefing probably would be given by   (who works for AI).

Focus of briefing is substantive rundown to include the Phenix program. To get Andy familiar with the activity, the need to monitor, and an item of interest/concern from a DOD perspective. Why? Among other things, because Andy may have some money he would be willing to allocate to keep the contract DIA has had going -- the contratct that allowed for those unique contacts and sourcing leads. I have the impression that Lin Wells wants to urge Andy to contribute if we can "interest" him.

Please coord among yourselves as to who will set up and plse include me. I think we need no more than 3 from PC   but that is their call. This should be done ASAP, because Lin is orchestrating interest in this subject in OSD prior to the Yeltsin visit. Thanks, Jay

Approved For Release 2003/04/18 : CIA-RDP96-00789R002600030012-8

The above declassified CIA documents on "signal nonlocality" evidence were provided by Gary Bekkum.

Quantum Code Breaking?

The above CIA/DIA Star Gate program depended on empirical "signal nonlocality" of conscious matter in strong violation of micro-quantum theory that is the basis of quantum cryptography described below. Iran is capable of getting there ahead of USA, so are many smaller groups. See Sir Martin Rees "Our Final Hour."

"Electronic commerce in its current form is saved from a catastrophic collapse only because the construction of a full-size quantum computer is, for the moment, eluding our technological capabilities. And we can only shiver to think of the effect that such a collapse of classical cryptography could have on national security."
Nature Physics 1, 2-4 (2005) ☐ Is information the key? Gilles Brassard

Why quantum computing may be a pipe dream because "More is different" (P.W. Anderson):

"The generation of heat is fundamental to [computational] reliability. ... the decisions that count are irreversible ... the quantum computer (is) a fundamentally new kind of computational hardware that would exploit the entanglement of the quantum wave function to perform calculations presently impossible with classical computers ... The impossibility of factoring a number that is the product of two large primes in reasonable time with conventional computers is the basis of modern cryptography. However, quantum computation has a terrible Achilles Heel that becomes clear when one confronts the problem of reading out the answer: the effects that distinguish quantum computers from conventional ones also cause quantum indeterminism. Quantum mechanical wave functions do indeed evolve deterministically, but the process of turning them into signals people can read generate errors ... the frenzy over quantum computing misses the key point that the physical basis of computational reliability is emergent Newtonianess ... The view that this problem is trivial is a fantasy spun out of reductionist beliefs ... Naturally, I hope I am wrong, and I wish those who invest in quantum computing the best of luck. I also ask that anyone interested in investing in a bridge in lower Manhattan contact me right away" Robert B. Laughlin, Nobel Prize in Physics, Professor at Stanford University, "A Different Universe" pp 59-65

Gilles Brassard is in the Département d'informatique et de recherche opérationnelle, Université de Montréal, Québec H3C 3J7, Canada.

"Quantum information science has brought us novel means of calculation and communication. But could its theorems hold the key to understanding the quantum world at its most profound level? Do the truly fundamental laws of nature concern — not waves and particles — but information? Imagine, what if all of quantum mechanics could be derived simply by taking those two quantum cryptographic theorems as axioms? ... But, even if the quantum computer existed, could it perform calculations that are impossible in the classical world? ... Quantum computing was at first regarded as a mere theoretical concept, but interest in it grew when Peter Shor discovered a way to use its capabilities to factorize large numbers efficiently. Such a computer would threaten the public-key

cryptographic schemes currently in use, in particular for the secure transmission of credit card numbers over the Internet. Electronic commerce in its current form is saved from a catastrophic collapse only because the construction of a full-size quantum computer is, for the moment, eluding our technological capabilities. ...

Enter cryptography
This begs the question: are there information-processing tasks that are impossible even in principle in the classical world, but that become possible through quantum mechanics? Even though unpublished for nearly fifteen years, the answer came to Stephen Wiesner well before anyone had thought of quantum computing. Around 1970, he discovered that quantum-mechanical effects could be used to produce banknotes that would be impossible to counterfeit. Because quantum information cannot be cloned, Wiesner realized that a banknote that contained quantum information would be impossible to copy. .... For ages, mathematicians had searched for a system that would allow two people to exchange messages in absolute secrecy. In the 1940s, Claude Shannon proved that this goal is impossible unless the two communicating parties share a random secret key that is as long as the message they want to communicate; moreover, that secret key can be used once only. In quantum cryptography, however, this pessimistic theorem can be thwarted by exploiting both the impossibility of measuring quantum information reliably and the unavoidable disturbance caused by such measurements. When information is appropriately encoded as quantum states, any attempt by an eavesdropper to access it necessarily entails a probability of spoiling it irreversibly. This disturbance can be detected by the legitimate users, allowing them to establish an unconditionally secure confidential channel with no need for a shared secret key. After we reported the first experimental realization of quantum cryptography, David Deutsch wrote in New Scientist: 'Alan Turing's theoretical model is the basis of all computers. Now, for the first time, its capabilities have been exceeded. It is interesting to note that quantum computers threaten most of the classical cryptographic schemes in use today, but that quantum cryptography offers an unconditionally secure alternative.'"

Only if the perfect no-cloning theorem prevents "signal nonlocality" as defined in papers by Antony Valentini now at the Perimeter Institute. If micro-quantum theory is to macro-quantum theory (with hidden symmetries in the ground state of large systems) as special relativity is to general relativity, then the "unconditionally secure alternative" could be the "Maginot Line" of the National Security Corporate State. The Fat Lady has not sung on this yet and policy wonks in USG Intelligence should not be lulled into a false sense of security by the above kinds of statements.

"The fact that no manipulations taking place at some point in space can have an instantaneously observable effect at some remote other point (the 'no-signalling property'); and that information cannot be cloned."

Don't be so sure. However, one can see that factual violation of the 'no-signalling property" brings the whole quantum cryptography program down like an unstable house of cards. Quantum security rests on shaky ground that could turn into quicksand. Don't bet on this horse. Not yet at least.

How Thoughts Shape Matter & Vice Versa

Stuart Hameroff had a good idea that the sub-neuronal nano-scale microtubules are the Seat of the Soul. Then he went and ruined it by collaborating with Roger Penrose, who though a genius of the highest rank, got all muddled in the Copenhagen fairy tale of "collapse of the state" - really a remnant of Marxist fantasy as effective in physics as it was in economics and politics.

On Apr 30, 2004, at 9:43 AM, Brian Josephson wrote:

"Dear Jack,

How you describe flashes on a screen as particles hit it, if not by collapse? Or are you a Many Worlds Theory person?"

The same way Basil Hiley does. For electrons they are real localized IT particles, like Wheeler's [micro] geons made out of the emergent micro g_{uv} field, but with strong short-range gravity, piloted by the BIT quantum potential.

Photons are trickier and they do have a kind of superluminal collapse as described in the book "The Undivided Universe." But what collapses is not the super-quantum BIT potential, but the energy-carrying IT classical EM field configuration, as I recall. So this is a very different idea of collapse. Basil will correct me here if I misremember?

On May 1, 2004, at 2:08 AM, Basil Hiley wrote:
"This is a succinct way of getting the main idea across. But see chapter 11 in the Undivided Universe or Phys. Rep. 144, 349-75, 1987 for a more technical discussion. In both places we outline the principles involved using a scalar field only to keep the mathematical symbolism as simple as possible. The extension to the em field was discussed by Pan Kaloyerou and presented in Phys. Rep. 244, 287-385 (1994)."

The "many worlds" are the basins of attraction for the IT "extra variables" on the BIT pilot wave landscape. They can be empty of real IT particles, but they are still physically real in the sense of inactive/active qubits. That's for micro-quantum theory. In macro-quantum theory the coherent order parameter combines IT with BIT in a single "two-way relation" or "back-action" in the sense of Bohm and Hiley's p. 30 of "The Undivided Universe." The quantum BIT wave now has an IT source unlike micro-quantum theory. The nonlinearity and loss of the projective ray property of Dirac's ket, at this emergent level of new orders of complexity mean spontaneous self-organization with presponse signal-nonlocality, seen for example in Rupert Sheldrake's "animal telepathy" data, in Dick Bierman's presponse data and in Puthoff and Targ's remote viewing data, because we are no longer constrained by the old rules. The idea of the ensemble and Born probabilistic "sub-quantum equilibrium" is gone. Stapp's "orthodox" paradigm is gone, however his nonlinear model of "retro-psychokinesis" is needed. The algorithm itself has mutated, It has adapted to the new conditions of "More is different."

(Dec 24 12:23:38 PST, 1996)

Henry Stapp opens his book, Matter, Mind, and Quantum Mechanics with the remark that the world divides into "thoughtlike things and rocklike things." Stapp also notes that William James at the close of the 19th Century pointed to the fact that there was no way in the classical physics of the time to formulate how a thought could influence the brain. This led to Watson's classical materialist behaviorism based on Pavlov, to Crick's "Astonishing Hypothesis" and to the current ill-conceived classical theories of consciousness of Churchland, Dennett, Calvin, Minsky, Edelman and others too numerous to mention. :-)

The orthodox Copenhagen interpretation of Niels Bohr is totally idealistic asserting that the world at the fundamental level of quantum reality only consists of thoughtlike things, which then "collapse" into the rocklike things of classical physics. There are several variations on this idea due to Heisenberg, von Neumann, Wigner, Wheeler and Penrose, but they all involve two qualitatively different kinds of dynamics. One is the "unitary" deterministic time evolution of the thoughtlike quantum wave pattern of an isolated system between measurements by an external system. The second is the "non-unitary" indeterministic "collapse" of the pattern to an actual classical "rocklike" thing.

David Bohm, directly under Einstein's influence at Princeton in 1951 or so, had an epiphany after writing his Copenhagen-based book, Quantum theory, and realized that quantum reality has both thoughtlike and rocklike things. Bohr said it was impossible to visualize the motion of a tiny rocklike particle in quantum reality. Bohm proved Bohr wrong. Bohm's thoughtlike thing is the quantum "pilot wave." His rocklike thing is the "hidden variable." That was a poor choice of terminology, and I will use John Bell's later word "beable" instead of "hidden variable." In Bohm's version the beable is the actual position of the particle at a moment of time for particle mechanics. The beable is also the entire classical field configuration over all space at a moment in time in quantum field theory. The quantum wave properties of particles are caused by a new kind of nonlocal form-dependent "quantum potential." This is how thought shapes brain matter. The quantum particle properties of classical fields are caused by a similar "super-quantum potential." The photon is an example of the latter. There is a very important qualitative difference between Henry Stapp's quantum theory of the mind-brain based on an ontological collapse and my post-quantum theory of the mind-brain based upon Bohm's idea of "active information." So let's see what Bohm means by "active information." The following quotes are from 'The Undivided Universe" with Basil Hiley.

"The electron actually is a particle with a well defined position x(t) which varies continuously and is causally determined." P. 29

Note the path x(t) of the rocklike particle is "causally determined" in quantum mechanics. In contrast, with post-quantum back-action (i.e., the Holy Spirit, the Breath of God, to use qabalistic images) the particle, now attached to its thoughtlike quantum pilot wave, becomes alive and participates in its self-determination even opposing non-self Darwinian natural selection pressures in some "robust" cases.

"This particle is never separate from a new type of quantum field that fundamentally affects it ... it too changes continuously and is causally determined." P. 29

Note use of the word "never." "What never? Well, hardly ever." (HMS Pinafore). If it were really "never", then that settles it. There is no possibility of personal immortality and Christ's message would then be false. However, recent research at IBM on "quantum teleportation" suggests that the thoughtlike quantum pilot wave pattern of implicate information can be teleported between different rocklike particles. If so, then one brain could experience the actual experience of another brain distant from it in space and time. Note also, as for the particle, the causal determination of the wave transforms to living self-determination when post-quantum back-action takes over from random environmental "decoherence." The rocklike particle and its attached thoughtlike pilot wave feedback error-correcting information back and forth between each other to become a self-determining conscious post-quantum adaptive system. The important point here is to realize that this is not possible in quantum mechanics. Bohm explains why it is not possible right here:

"Finally it should be pointed out that unlike what happens with Maxwell's equations for example, the Schrodinger equation for the quantum field does not have sources, nor does it have any other way by which the field could be directly affected by the conditions of the particles." P. 30

The above remark is the most important sentence in Bohm's entire book.

This is the seed out of which the post-quantum physics of the 21st Century will blossom. What Bohm is saying here is that quantum mechanics has zero back-action. Back-action means just the opposite. The post-quantum field (or pilot wave) does have sources and is directly affected by the conditions of the particles. Thus, there is a two-way relationship between the quantum wave and the classical particle or beable in post-quantum physics. In contrast, there is only a one-way force of quantum wave on classical particle in quantum physics. Eberhard's theorem of quantum physics has a corollary that says that all paranormal phenomena based on clairvoyant telepathic faster-than-light and precognitive backward-in-time communication using nonlocal connectivity is impossible. Eberhard's theorem does not work in post-quantum physics. Note that the super-quantum fields of the classical gauge force fields are also directly affected by the conditions of the latter's configurations over all space at a fixed moment of time. This Hamiltonian picture must be extended to the Feynman path Lagrangian picture. Just as special relativity with zero space-time curvature is extended to general relativity with non-zero space-time curvature to explain gravitation, so I have, single-handedly in the face of enormous opposition from my peers, extended quantum mechanics to post-quantum mechanics to explain consciousness. I must thank Nick Herbert for pushing me into this strategy with his insistence that I make a new "Sarfatti Mechanics." Nick Herbert's 1982 FLASH Gedankenexperiment attempting signal nonlocality in micro-quantum theory failed but it launched modern quantum entanglement information theory because it triggered the "no-cloning theorem" that even Eugene Wigner missed. You cannot perfectly clone a quantum state without violating the linearity and the unitarity (conservation of probabilities for alternative outcomes of a quantum measurement) of micro-quantum theory. However, you can imperfectly clone and this is discussed in "Quantum Cloning" in the October 2005 Reviews of Modern Physics. Macroquantum theory, in contrast is neither linear nor unitary and may permit the "signal nonlocality" that is forbidden by the no perfect cloning theorem of micro-quantum theory.

Bohm continues:

"This of course constitutes an important difference between quantum fields and other fields that have thus far been used."

Bohm's "important difference" is missed not only by laymen, but also by theoretical physicists who should know better. The difference is that the classical gauge fields, like Maxwell's electromagnetic field, are local in space-time. In contrast, the quantum field of n point particles is local in a higher 3n dimensional classical configuration space C_{3n} at each instant of time[xl] in the Hamiltonian picture. Therefore, it is objectively nonlocal in the ordinary 3D space of the classical gauge field. Similarly, the super-quantum field of the classical gauge field is local in the infinite-dimensional classical configuration space $C_{infinity}$ because the classical gauge field is the limiting case of n coupled particles with n approaching infinity as a discrete cellular space lattice approaches the classical continuum of real numbers. The classical Wheeler superspace for the three-dimensional geometries of the metric gravitational gauge field of general relativity is an example of a $C_{infinity}$ classical rocklike configuration space. Note that classical configuration spaces are not the same as the quantum Hilbert spaces of pilot wave functions. The classical configuration space is the rocklike space in which the beable moves. The quantum Hilbert space is the thoughtlike space in which the pilot wave moves. The thoughtlike Hilbert space is a fiber space, which has the rocklike classical configuration space as its base space. Thus, the thoughtlike quantum Hilbert space of pilot waves of "funda-MENTAL"[xli] (e.g., the "Chalmers field") implicate information and the rocklike classical configuration space of the beable, form the "mind-brain fiber bundle" when the post-quantum back-action is not zero. This is how the post-quantum physics of consciousness connects to the mathematics of Saul Paul Sirag in the latest edition of Jeffrey Mishlove's The Roots of Consciousness. Sirag was the vice-president of the original Esalen Physics Consciousness Research Group when I was president in the mid-seventies. Like Zukav's "The Dancing Wu Masters", so too the new edition of "The Roots of Consciousness" deletes much of the material I wrote for the first edition of both of those books. Not a very smart move on the parts of the authors! (Grin and Sneer) Now Bohm shows how quantum mechanics is the approximate limiting case of zero back-action just like the similar case of the globally flat space-time of special relativity, which has zero gravitational field in it.

"As we shall see, however, the quantum theory can be understood completely in terms of the assumption that the quantum field has no sources or other forms of dependence on the particles. We shall in chapter 14, section 14.6, go into what it would mean to have such a dependence and we shall see that this would imply that the quantum theory is an approximation with a limited domain of validity." P.30

Going to 14.6 P. 346 we find Bohm asserting:

"One way to make the Schrodinger equation dependent on the particle positions (so that there would be a two-way relationship[xlii] between wave and particle) can be seen by considering equation 14.1 ... From the same arguments as apply to the GRW approach, it would follow that the overall wave function would tend to 'collapse' towards the actual particle positions, so that, in a large scale system, the empty wave packets of our interpretation would tend to disappear."

Stapp interprets Bohm's above words "so that, in a large-scale system, the empty wave packets of our interpretation would tend to disappear" to show the equivalence between his ontological post-Heisenberg-James collapse theory of consciousness and my back-action theory. If so, my back-action theory is more detailed and can explain the self-determination of free-will choices in individual events, which he explicitly admits he cannot explain. In any case, it is clear that Stapp is proposing a post-quantum theory of consciousness not a quantum theory as he sometimes seems to suggest.

Now let's return to the main intent of this article which is Bohm's notion of "active information" bearing in mind that the key difference right now between Stapp's theory and my theory is whether or not the "inactive information" of the post-quantum mind-brain conscious adaptive system is wiped away moment to moment as our streams of inner felt-consciousness, like "Ol Man River just keeps on flowin along." The interaction of the classical particle system with its quantum pilot wave field (i.e., "quantum potential" also called "Q") is qualitatively different from the interaction of a classical charge with a classical local electromagnetic field. This is a key point that many paraphysicists completely miss in tying to model the paranormal. Since the CIA and the DOD of the USA have paid good money for such models, what I am saying here is important to the taxpayer. (Grin) Bohm writes:

"For the quantum potential has a number of strikingly new features which do not cohere with what is generally accepted as the essential structure of classical physics. .... The first of these new properties can be seen by noting that the quantum potential is not changed when we multiply the field psi by an arbitrary constant. (This is because psi appears both in the numerator and the denominator of Q) This means that the effect of the quantum potential is independent of the strength (i.e. the intensity) of the quantum field but depends only on its form. By contrast, classical waves, which act mechanically (i.e., to transfer energy and momentum, for example, to push a floating object), always produce effects that are more or less proportional to the strength of the wave. For example one may consider a water wave, which causes a cork to bob. The further the cork is from the centre of the wave the less it will move. But with the quantum field, it is as if the cork could bob with full strength even far from the source of the wave." P. 31

Bohm then compares this to:

"A ship on automatic pilot being guided by radio waves. Here, too, the effect of the radio waves is independent of their intensity and depends only on their form. The essential point is that the ship is moving with its own energy, and that the form of the

radio waves is taken up to direct the much greater energy of the shiP. We may therefore propose that an electron too moves under its own energy, and that the form of the quantum wave directs the energy of the electron." P. 32

Clearly the "form" dependence of the quantum motion is thoughtlike as when we do something because we intend to do it. The word "form" is identical to the way I used "context" as in "context-dependence" in the past. Bohm goes on to say that the new form-dependence implies that a particle may not move uniformly in a straight line in the absence of all classical forces. That is Newton's laws are modified in quantum reality.

"Moreover, since the effect of the wave does not necessarily fall off with the distance, even remote features of the environment can profoundly affect the movement." P. 32

"Effects of this kind are indeed frequently encountered in ordinary experience wherever we are dealing with information. ... We explain the interference properties by saying that the quantum field contains information, for example about the slits, and that this information is taken up in the movements of the particle..." P. 35

Note Bohm's use of the words "is taken up in the movement" This is obviously how thoughts influence brains. I mean it is really obvious. We see how the brain sucks up the messages from the mind.

"We have in this way introduced a concept that is new in the context of physics -- a concept that we shall call 'active information'... The basic idea of active information is that a form having very little energy enters into and directs a much greater energy. The activity of the latter is in this way given a form similar to that of the smaller energy."

A Quantum Loop Hole in the Second Law of Thermodynamics?

$$\frac{W_{out}}{Q_{in}} \leq 1 - \frac{T_{cold}}{T_{hot}}$$

$$\frac{T_{cold}}{T_{hot}} < 0$$

$$\frac{W_{out}}{Q_{in}} > 1$$

Use a hot negative qubit temperature heat reservoir coupled to a cold positive temperature reservoir. All negative quantum temperatures are "hotter" than any positive temperature. Therefore, you can have over 100% efficiency, i.e. "over-unity" since energy can be sucked out of *both* heat reservoirs. This is not a violation of the second or the first laws of thermodynamics. It is a consequence of it when coupled with quantum theory. Note, however, that this formal argument does not demand that such a strange phenomenon happens. It only alerts us that if we observe it, it is a quantum loophole in the classical second law of thermodynamics not a violation of it nor of total energy conservation that is the first law of thermodynamics.

A good example of Bohm's active information is in the quantum Carnot engine that I invented as a thought experiment. Couple a very hot negative quantum temperature to a cold positive classical temperature and you will see active information at work in a very dramatic counter-intuitive way that is relevant to Herbert Frohlich's Bose- Einstein condensates in living microtubules. By "form", Bohm means a new kind of implicate quantum information of unique complex individuals that is qualitatively different from Claude Shannon's thermodynamic entropy-based explicate classical statistical information of actuarial ensembles of simplistically abstracted so-called identical individuals

"It is important to distinguish our concept of active information from the more technical definition of information commonly adopted in physics in terms of, for example, Shannon's ideas implying that there is a quantitative measure of information that represents the way in which the state of a system is uncertain to us ... such concepts have been used to calculate the objective properties of systems in thermodynamics and even black holes ... but we wish to propose here a quite different notion of information that is not essentially related to our own knowledge or lack of it. ... it will be information that is relevant to determining the movement of the electron itself." P. 35

Yes, indeed, just like the thoughts in your mind that determines the movements of the muscles in your body.

"What is crucial here is that we are calling attention to the literal meaning of the word, i.e., to in-form, which is actively to put form into something or imbue something

with form." P. 35

This is how the quantum mind, which is a pilot wave imbues the classical brain to which it is attached with coordinated behavior. The basic postulate here is that quantum waves are fundamentally thoughtlike. The universe naturally divides into thought and matter. Bohm never quite escapes Bohr's idealism that thought is more fundamental than matter. After all Bohm did write the best book on Bohr's version of quantum physics before Einstein turned him. Thus, Bohm till his dying day suspected that the thoughtlike Hilbert space of quantum waves generated material rocklike things. This would take us back to Stapp's theory but with more details. If Bohr is fundamentally right, then we would have beings like Q in Star Trek (TNG) materializing objects out of thin air rather than simply moving them psychokinetically at a distance because the pilot wave spreads out much further than the parts of its localized beables in the case of particles. We will have to wait and see. Meantime let us continue.

Getting back to the idea that active information does not use its own generally tiny energy, but uses the larger classical energy of its attached beable. Bohm writes:

"The sound energy we hear in the radio does not come directly from the radio wave itself, which is too weak to be detected by our senses. It comes from the power plug or batteries, which provide an essentially unformed energy that can be given form (i.e., in-formed) by the pattern carried by the radio wave. This process is evidently entirely objective and has nothing to do with our knowing the details of how this happens." P. 36

Here Bohm is attacking Bohr and Heisenberg's idea that quantum waves are only epistemological waves of knowledge in some mystical sense.

"The information in the radio wave is potentially active everywhere, but it is actually active, only where and when it can give form to the electrical energy, which, in this case is in the radio."

Bohm explains the computer this way as well.

"The information content in a silicon chip can determine a whole range of potential activities which may be actualized by giving form to the electrical energy coming from a power source. Which of these potentialities will be actualized in a given case depends on a wider context and the responses of a computer operator." P. 36

So far Bohm is discussing only classical information machines where both the informer and the informed are classical devices. The difference in the mind-brain system is that the mind is a quantum device and the brain is a classical device. So rather than a classical-classical information machine we now have a quantum-classical information machine. Furthermore, the two-way relation between the thoughtlike quantum wave and the rocklike classical brain beable means that each of them are simultaneously working as informer and informed. This sets up a self-referential Godel loop that Douglas Hofstadter calls a "strange loop" and Eric Harth calls a "creative loop." Consciousness requires such a post-quantum self-determining adaptive loop on the edge between classical determinism and quantum indeterminism. The capacity of Godel loops to jump out of their own systems into strange territory where none have gone before is the nonalgorithmic character of human understanding that Roger Penrose describes in his book, "Shadows of the Mind." What we have here is the solution changing its own generating equation in a globally self-consistent loop connecting all that was, all that is, and all that is to come. Bohm discusses our own DNA genetic code where we get closer to the quantum-classical information machine, i.e.

"The function of the DNA molecule. The DNA is said to constitute a code, that is to say, a language. The form of the DNA molecule is considered as information content for this code, while the 'meaning' is expressed in terms of various processes; e.g. those involving RNA molecules, which 'read' the DNA code, and carry out the protein-making activities that are implied by particular sections of the DNA molecule. ... in the process of cell growth it is only the form of the DNA molecule that counts, while the energy is supplied by the rest of the cell (and indeed ultimately by the environment as a whole). Moreover, at any moment, only a part of the DNA molecule is being 'read' and giving rise to activity."

This should put to rest the common confused New Age cliché that all energy is conscious. Most energy is uninformed classical localized energy. Only a small fraction of the energy in the universe is in-formed making it thoughtlike because of its nonlocal quantum character. But even this thoughtlike energy is not yet conscious because of the one-way relation where it acts on its beable, but is not acted back upon by its beable. Without this active two-way feedback command-control-communication there cannot be any coherent changes in the in-formed energy. It is these coherent changes, which are our inner felt-experiences of "qualia" that the psychologists of consciousness like to bandy about. Without this post-quantum two-way relation of back-action, so that the beable becomes a source of information for its attached pilot wave, there is only the uncontrollable noise of quantum randomness. So, God does not play dice with the classical universe. God does play fair dice with the quantum universe, but God loads the dice with consciousness in the post-quantum universe. This is like three Chinese nested boxes with the classical universe inside the quantum universe, and the quantum universe inside the post-quantum universe where it then halts! But consciousness is subjective and Bohm appears to be intent upon objectivity. However, he writes:

"While we are bringing out above the objective aspects of information, we do not intend to deny its importance in subjective human experience. ... even in this domain, the notion of active information still applies. A simple example is to be found in reading a map ... we apprehend the information content of this map through our own mental energy. And by a whole set of virtual or potential activities in the imagination, we can see the possible significance of this map. Thus the information is immediately active in arousing the imagination, but this activity is still evidently inward within the brain and nervous system. .. If we are actually traveling ... then, at any moment, some particular aspect may be further actualized ... even the information held by human beings is, in general active rather than passive, not merely reflecting something outside itself but

actually capable of participating in the thing to which it refers."

Again the idea of a loop appears. But now Bohm says something interesting. He says that Shannon's notion of information is "passive."

"Passive information may in fact be regarded as a limiting case in which we abstract from the activity of information. This is essentially the kind of information that is currently used in information theory, e.g. as used by Shannon. The puzzle in this approach is that of how information that is merely passive within us is able to determine actual objective processes outside of us. We suggest that passive information is rather like a map reflecting something of these processes which can guide us to organize them conveniently for our use, e.g. by means of algorithms that enable us to calculate entropy." P. 37

So the algorithmic nature of information, which Gell-Mann in his "The Quark and The Jaguar" never escapes, thus making his Santa Fe Institute a modern day Laputan Academy (Gulliver's Travels), is the fragmentary result of artificially abstracting or squeezing out all of the active nonalgorithmic understanding that Penrose has emphasized incurring the Wrath of the Laputans. (Grin and Sneer)

"If the notion of active information applies both objectively and subjectively, it may well be that all information is at least potentially active and that complete passivity is never more than an abstraction valid in certain limited circumstances." p 37

So far these remarks of Bohm are metaphoric setting the stage. He then goes to Feynman's "central mystery of quantum mechanics" the deceptively simple, yet paradoxical, double slit interference experiment in the low intensity limit so only one particle moves through the system at any one moment.

"We could say that this particle has the ability to do work. This ability is released by the active information in the quantum field, which is measured by the quantum potential. As the particle reaches certain points in front of the slits, it is 'in-formed' to accelerate or decelerate accordingly, sometimes quite violently." P. 37

Newton's second law of motion for the beable that force equals mass times acceleration is modifed in Bohm's theory to include the new nonlocal form-dependent quantum force in addition to the local intensity-dependent classical force. While the classical force pushes and pulls the way you learned physics in high school or college, the quantum force is qualitatively different. Remember, none of this can even be conceived of in Bohr's Copenhagen interpretation or Stapp's post-Heisenberg-James interpretation. This is a unique new physics only found in Bohm's interpretation. Bohm writes:

"Although [the modified Newtonian second law] may look like a classical law implying a pushing or a pulling by the quantum potential, this would not be understandable because a very weak field can produce the full effect, which depends only on the form of the [quantum] wave. We therefore emphasize that the quantum field is not

pushing or pulling the particle [classically] mechanically, any more than the radio wave is pushing or pulling the ship that it guides. So the ability to do work does not originate in the quantum field, but must have some other origin ... The fact that the particle is moving under its own energy, but being guided by the information in the quantum field, suggests that the electron or any other elementary particle has a complex and subtle inner structure (e.g. perhaps even comparable to that of a radio). This notion goes against the whole tradition of modern physics, which assumes that as we analyse matter into smaller and smaller parts its behavior always grows more and more elementary." P. 37

Bohm then compares his conjectured "inner complexity" of the individual electron to that of people. He notes that complex people obey simple statistical laws in crowds (i.e. ensembles) that are reproducible enough for insurance companies and advertising agencies to make lots of money. Getting back to the idea that "the particle is moving under its own energy." Bohm adds that: "the vacuum is ... full with an immense energy of fluctuation, revealed for example in the Casimir effect, it may be further suggested that ultimately the energy of this particle comes from that source.' P. 38

"The quantum information field may also have some energy. However, ... this must be negligible in comparison to the energy of the particle, which it guides ... A very important further implication of the notion of active information is that in a certain sense an entire experiment has to be regarded as a single undivided whole."

The mind is a large thing immune to environmental decoherence.

"To summarize: The formalisms that describe the holographic process and those that describe quanta of information apparently DO extend to scales other than the quantum. Today we use quantum holography to produce images with the technique of functional Magnetic Resonance (fMRI). The quantities described by terms of the formalisms such as Planck's constant will, of course, vary but the formulations will to a large extent be self-similar. The important philosophical implications for the brain/mind issue have been addressed in depth by Henry Stapp on several occasions (e.g 2003, —The Mindful Universe") as well as by many others including myself (e.g. Pribram, 1997, What is mind that the brain may order it?). "

Stapp's theory is no good because he does not ask the right question. Stapp uses the collapse idea and for that reason throws the consciousness baby out with the bath water out into the briar patch filled with thorns and poison oak in the not hallowed but hollow withered dead Halls of Ivy. Even though von Neumann, London, Wigner and Penrose played with the idea it is no good and does not work. A reading of P.W. Anderson's book "A Career in Theoretical Physics" shows why the quantum measurement approach is no good for consciousness studies. Almost all the quantum theory papers at the Tucson Consciousness Conferences are not correct. Henry Stapp came closest in his retro-PK paper in Phys Rev A for which he was roasted alive in Physics Today and other boring venues of academic respectability. Stapp then had a failure of nerve, and like Galileo before the Inquisition of the Immaculate Deception, recanted and has been under virtual house arrest ever since.

Quantum Brain Dynamics & Signal Nonlocality

"What everybody does seem to agree on is that the use of this effect is limited. You can't use it to send a message, for example. Leonard Susskind, a Stanford theoretical physicist, who called these entanglement experiments "beautiful and surprising," said the term 'spooky action at a distance,' was misleading because it implied the instantaneous sending of signals. "No competent physicist thinks that entanglement allows this kind of nonlocality." By Dennis Overbye, "Quantum Trickery: Testing Einstein's Strangest Theory" New York Times, December 27, 2005

Yes, for orthodox quantum theory, but Antony Valentini, a competent physicist now at the Perimeter Institute, showed the limits of orthodox quantum theory with signal locality and no-cloning and one can construct more general theories or post-quantum theories with signal nonlocality that seem to describe living matter because of the brain "presponse" experiments of Ben Libet, Dean Radin and Dick Bierman. Also because of the CIA experiments in remote viewing by Hal Puthoff and Russell Targ at SRI in the 1970's. Mainstream physicists reject this data for bad reasons. Brian Josephson was one of the first to notice this as was Roger Penrose. The post quantum theories are nonlinear and nonunitary and the Born probability rules break down completely. Post-quantum theories (Weinberg & Stapp played with them) are to orthodox linear unitary micro quantum theory of simple systems with locally random individual events as general relativity is to special relativity. Irreducible randomness is analogous to global flatness. Human consciousness represents a strong violation of orthodox quantum theory if one believes the data on remote viewing for example. What makes the difference here is ODLRO in open pumped complex systems not in thermal equilibrium. ODLRO comes from a spontaneous breakdown of symmetry in the ground state of a large system. The result is a manifold of degenerate ground states of the same minimum energy on a landscape of possible solutions to the field equations. The topology of this manifold determines the essential physics from the resulting stable non-trivial homotopy "defects." This is part of the program of P.W. Anderson's "More is different" as a general physics for the emergence of new orders in complex systems. This theory explains both gravity and consciousness as spontaneously self-organizing emergent phenomena from a simpler substratum.

"Henry Stapp in two excellent articles (Stapp 1997a and b) reviews the development of quantum theory and outlines how it is essential to understanding the mind/brain relationship. Stapp sets up the issue as follows. —Brain process is essentially a search process: the brain, conditioned by earlier experience, searches for a satisfactory response to the new situation that the organism faces. It is reasonable to suppose that a satisfactory response will be programmed by a template for action that will be implemented by a carefully tuned pattern of firings of some collection of neurons." Karl Pribram

The neurons are too coarse-grained to explain the hard problem of the emergence of inner consciousness - of who we really are.

"The executive pattern would be a quasi-stable vibration that would commandeer certain energy resources, and then dissipate its energy into the initiation of the action that it represents." Karl Pribram

OK for our machine zombie levels of behavior, but it does not touch on the hard problem

"Stapp further notes that —the relative timing of the impulses moving along the various neurons, or groups of neurons, will have to conform to certain ideals to within very fine levels of tolerance. How does the hot, wet brain, which is being buffeted around by all sorts of thermal and chaotic disturbances find its way to such a tiny region in a timely manner?" Karl Pribram

"The Question is: What is The Question?" John A. Wheeler

Giant Psi waves are immune to heat decoherence.

"Further: —How in 3n dimensional space (where n represents some huge number of degrees of freedom of the brain) does a point that is moving in a potential well that blocks out those brain states that are not good solutions to the problem --- but does not block the way to good solutions find its way in a short time to a good solution under chaotic initial conditions?" Karl Pribram

Signal nonlocality - presponse precognition is essential. This violated quantum theory. Precognition certainly aids in solving a search problem and in pattern recognition. Precognition is delayed choice one can use, not the orthodox one Wheeler discusses that we cannot use to survive.

"Stapp notes that classical solutions to this problem won't work and that —the quantum system [will work as it] has the advantage of being able to explore simultaneously (because the quantum state corresponds to a superposition of) all allowed possibilities." Karl Pribram

You need creative tension, some linear superposition, but also enough nonlinearity, in sense of the Landau-Ginzburg equation for the giant quantum Psi field, to have signal nonlocality between different parts of the brain that get their marching orders from the same Psi mind field.

"Stapp provides a viable metaphor in a glob or cloud of water acting together rather than as a collection of independently moving droplets. —The motion of each point in the cloud is influenced by its neighbors." Karl Pribram

For that we have the nonlocal Bohm quantum potential of the giant Psi wave. But now we have signal nonlocality not signal locality. The linear unitary Schrodinger equation of micro-quantum theory in huge configuration space is no good. We need the nonlinear nonunitary Landau-Ginzburg equation in ordinary space. Giant quantum mind fields are not "projective rays". "More is different." Trash all of Von Neumann's "Quantum Theory of Measurement." It's useless for the "hard problem." Everything you read in the New Age pop books on the physics of consciousness is "not even wrong." The rules of the Glass Bead Game have changed qualitatively. There are more things between Heaven and Earth than are dreamt of in the Copehagen "philofawzy." Something is rotten in the "collapse of the state" in Denmark.

"However classical holography will also do just this." Karl Pribram

Giant quantum wave mind fields are like classical holograms in some respects.

"But the advantage of holonomy, that is quantum holography, is that it windows the holographic space providing a —cellular" phase space structure, in patches of dendritic fields thus enhancing the alternatives and speed with which the process can operate. In short, though the information within a patch is entangled, cooperative processing between patches can continue to cohere or de-coherence can —localize" the process." Karl Pribram

No, all the Tucson Pundits are completely confused. It's not entanglement they need, it's signal nonlocality that they need. They are, all of them, very far from asking the right question. Antony Valentini formulated what is close to the right question in his "sub-quantal non-equilibrium" AKA Landau-Ginzburg equation.

"With regard to evidence regarding the scale at which quantum processes are actually occurring, a number of publications have reported that quantum coherence characterizes the oscillations of ions within neural tissue channels." Karl Pribram

Macro-Quantum Immunity to Environmental Decoherence

"The question immediately arises as to whether decoherence occurs when the channels communicate with each other and if so, how. Stapp notes that —phase relationships, which are essential to interference phenomena, get diffused into the environment, and are difficult to retrieve." Karl Pribram

Stapp is wrong. He is not looking at the correct giant quantum wave that is the physical field of the living conscious mind and that is immune to hot wet thermal decoherence from its generalized phase rigidity that allows stable mental holograms to persist.

"These decoherence effects will have a tendency to reduce, in a system such as the brain, the distances over which the idea of a simple quantum system holds." Karl Pribram

P.W. Anderson's "generalized phase rigidity" of emergent complexity gives an immunity against environmental decoherence. The quantum phase is fragile, but the long range coherent macro-quantum phase of the living mind hologram is not fragile, it is "rigid" immune from ordinary sources of decoherence. Jorge Luis Borges, the Homer of our time, also a Qabalist would think of Bohm's landscape as all the books in The Library waiting to be read with active information as the consciousness elicited in the reading of a book, nay a page – depending on what adaptive wavelet resolution your mind decides on. It's curious that Lenny Susskind has a large-scale "cosmic landscape" where Bohm's hidden-variable theory has its most natural setting in which the IT hidden variable is a system point rolling on the landscape of the BIT pilot wave. Ordinary quantum theory with unitarity, no cloning and signal locality has the hidden variable system point as a passive test particle. The system point receives its marching orders from the gradients in the landscape of hills and valleys in "configuration space". The system point reacts, but it does not act. It's reaction without action. Post-quantum theory, with inner consciousness, I posit, is when the relation between IT particle system point and the intrinsically mental BIT pilot wave landscape is "two-way" in a self-creative adaptive spontaneously self-organizing feedback-control loop.

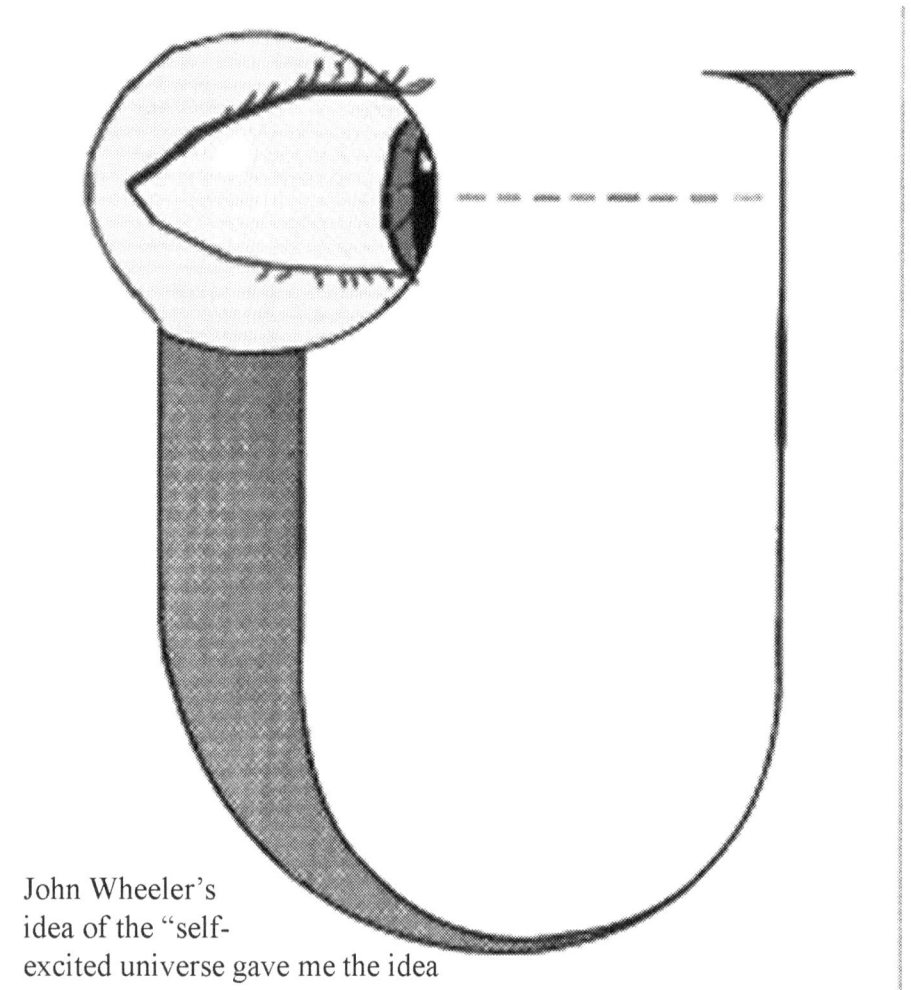

John Wheeler's
idea of the "self-
excited universe gave me the idea

IT FROM BIT + BIT FROM IT

Roy Lichtenstein's "Tear"

Limitless Mind in Super Cosmos

I had to omit the quote from Russell Targ's book of this title to keep the book of manageable length. The key physics idea behind remote viewing is "signal nonlocality" which allows messages to be locally decoded at the IT nodes of an entangled nonlocally connected network with a shared BIT thought pattern. This is not possible if micro-quantum theory is the final story of physical reality. Micro-quantum theory is only a limiting "equilibrium" case of the more general "post-quantum" or "macro-quantum" theory which allows "sub-quantum non-equilibrium" in which the usual Born probability interpretation breaks down as does the notion of unitary evolution of the thought BIT pattern of the network. The remote-viewing data that Russell talks about below in addition to the "presponse" experimental data of Dick Bierman and others strongly suggests that all life is macro-quantum with signal-nonlocality. This is in violation of the micro-quantum limiting case, which only works in simple structures like particle beams in scattering experiments or in other simple systems, generally in thermodynamic equilibrium or not far from it where the fluctuation-dissipation theorem works, like radiating gas in a star, where emergent "More is different" collective phenomena are either not present or not detected in the measurements. Just as special relativity is a limiting case of general relativity, so, also, is micro-quantum theory with signal locality a limiting case of macro-quantum theory with signal nonlocality. In both of these examples, the key principle is the generalized action-reaction principle. Both micro-quantum theory and special relativity violate the generalized action-reaction principle also known as "background independence." This means there is no nondynamical absolute that acts without being directly acted upon by what it is acting on. Bohm and Hiley call this the "two –way relation." As the "Physics Chanteuse" Lynda Williams told me, Paul Hewett formulates this as "You cannot touch without being touched. I call it "the two way street" or "It takes two to tango" principle, or "What comes around goes around" principle. General relativity and macro-quantum theory are both "background independent" restoring the generalized action-reaction principle. This means that they are "relational" in Rovelli's sense. Now, in this book, I show that Einstein's seemingly classical curved space-time of gravity is also really a macro-quantum phenomenon with inherent signal nonlocality essential for the generation of inner conscious experience of within a BIT thought field. Now Sir Roger Penrose of Oxford as intuited that quantum gravity and consciousness are linked. One stumbling bloc to that has been the assumption that the energy scale of quantum gravity is huge at 10^{19} Gev or 10^{-33} cm, however this is probably a very wrong idea because gravity is enormously strong at the scale of 1 fermi (10^{-13} cm) in order to keep electrons and quarks stable. Otherwise they would spontaneously explode under the self-repulsion of their spatially-extended electric charge distributions. They are not really point particles, certainly not in string theory with IR/UV duality. They only look like point particles because of the small-scale enormous space-time warping in high-energy scattering momentum transfers between probe and target since the effective micro-gravity coupling is 40 powers of ten stronger than Newton's large-scale gravity coupling on the scale of a fermi as Abdus Salam first suggested in the early seventies. This allows Wheeler's Geometrodynamics of "Mass without mass" etc. to actually work as a Bohmian model of elementary particles as "extra variables" in the sense of "Einstein's Vision" where the electrons et-al are solitons in the geometrodynamic field. You will see below in Russell's essay that Schrodinger believed in the universal cosmic consciousness. Hawking mentions "The Mind of God." In this book you will see what that really means. You will also see in Cabalist Carlo Suares's "Eight Propositions" the same idea and you see it again in Jacques Vallee's "Fastwalker" on the "consciousness of space." Since curved space-time with gravity is emergent in the sense of Andrei Sakharov as the phase wiggles in a local macro-quantum field, it is not surprising that curved space-time is endowed with a cosmic consciousness, or "Mind of God" because of signal nonlocality. It is counter-intuitive that nonlocal micro-quantum theory has signal locality or "passion at a distance" whilst local macro-quantum theory has signal nonlocality essential for inner consciousness in the physical BIT thought fields. There is no contradiction between signal nonlocality in the BIT thought field with the light cone causal structure in the IT geometrodynamic field. The cosmic BIT thought field spits out IT curved space-time in the wiggles of its coherent holographic phase. In spite of our local stupidity, we most likely live in an "Intelligent Universe" in Sir Fred Hoyle's sense with a conscious cosmic mind in Schrodinger's sense below.

"However, if we discover a complete theory, it should in time be understandable by everyone, not just by a few scientists. Then we shall all, philosophers, scientists and just ordinary people, be able to take part in the discussion of the question of why it is that we and the universe exist. If we find the answer to that, it would be the ultimate triumph of human reason -- for then we should know the mind of God." (p.193)

Stephen Hawking, "A Brief History of Time" (1996)

Satori

About 25 years ago Kim Burrafato and I saw Dennis Schmidt's novel only for a minute or two in a bookstore in the Marina district of San Francisco during a brief stop on a power walk. I quickly flipped through the pages and noticed a reference to me and another physicist as inventors of a warp drive. I called Kim over and showed it to him. I was not working on that problem at the time. I did not buy the book, but the strange synchronicity of seeing myself mentioned in a work of science fiction stuck in my mind. I remembered the incident but never found the quote in the book again until now (February, 2004).

On Feb 5, 2004, at 6:05 PM, Tony Smith wrote:

"The excerpts clearly show you as an advancer of knowledge, but it seems to me (maybe I am misreading, but I think that my quotes are pretty much in accurate context) that the quotes of statements by the Bishop about Heisenberg, Bohm, and Finkelstein are approving statements by the authoritarian Bishop, which would imply that they were considered to be on the side of repression rather than free research. For example, the Bishop said that they were 'against the iron rule of reason."

You found Dennis Schmidt's 1981 sci-fi book "Satori" great! That confirms I was not imagining it! Bohm too. Uh Oh! I better read that book. How did you get it? Online? It looks like I substituted "Brian Josephson" for "Alain Aspect" in my memory. Is Josephson mentioned in the book? Aspect did visit me in North Beach[xliii] with Lawry Chickering who was running the Reagan think tank ICS set up by Cap Weinberger and Ed Meese. We had a party for Alain Aspect with Lawry Chickering at Jagdish Mann's place on Lombard Street near Coit Tower. This is the time of the Chickering letter to Under Secretary of Defense Richard De Lauer that I wrote with Lawry that Cornell's N.D. Mermin (Physics Today), Gell-Mann (Quark and Jaguar) and Jeremy Bernstein (New Yorker) all wrote about. This was when we were participating in the formulation of the SDI strategy with Marshall Naify and Chickering. The story is in Destiny Matrix. Sounds like a takeoff on Dostoevsky's "Grand Inquisitor" scene in "The Brothers Karamazov" and a similar scene in Aldous Huxley's "Brave New World."

Just in time to put into my book "Super Cosmos." I did not know David Finkelstein was in it also? That means Schmidt must have read "The Dancing Wu Li Masters" or maybe he was an est person? As I say Kim Burrafato and I only looked at it a few seconds back in ~ 1981 in a book store on our exercise routine out to the Marina Green by the Golden Gate Bridge. What Arthur Koestler called a "Library Angel." I never read the book, nor do I own it.

"Heisenberg stayed in Nazi Germany, so I can see the author, Schmidt, having that opinion of him for political reasons, and also perhaps because Schmidt might have thought of Heisenberg's late efforts at a unified field theory as efforts a la Einstein to replace quantum probability with a classical-type field theory - do you recall any such thing?"

I knew Heisenberg slightly. I met him with Feynman at Cal Tech and I stayed at his Institute in Munich in July 1966 at a NATO institute with John Wheeler and Martin Kruskal. This was before I went to UKAERE Harwell to write the paper with Marshall Stoneham on spontaneous broken symmetry Goldstone - Jahn Teller Effect in Proc Phys Soc London. Heisenberg had a Landau-Ginzburg sort of Dirac eq as I dimly recall.

"Maybe the author Schmidt thought of Bohm as trying to bring classical order instead of probabilistic uncertainty to physics (note that would NOT apply to your post-Bohm quantum theory with back-reaction)."

I had no idea of "two-way" spontaneously self-organizing intrinsically conscious back-action with presponse signal nonlocality in 1981. I did not get that idea until 1994 reading Bohm and Hiley P. 30 of "The Undivided Universe" in "A Clean Well-Lighted Place for Books" at Opera Plaza in San Francisco. I had a girlfriend who owned a successful computer company who lived there. She had been quite a well-known mathematician before becoming an entrepreneur. Bohm and Hiley show that it is the absence of back-action in micro-quantum theory that is the reason for uncontrollable quantum randomness and signal locality forbidding the use of controlled entanglement as a direct stand-alone communication channel independent of classical signals.

"Do you have any idea why the author Schmidt might have put Finkelstein on the side of repression?"

You tell me. You were his student. I invited David Finkelstein to Esalen in 1976. I first met him with Lenny Susskind at Yeshiva University in the mid-60's when he (David) was trying to look like Ba Ba Ram Dass, so I figured he would be perfect for Esalen and fit right in, which he did in the late 70's. David then organized the Big Shot Werner Erhard Physics meetings with Stephen Hawking, Sidney Coleman, Feynman, ... Lenny Susskind mentions that on John Brockman's www.edge.org. I remember Sidney Coleman giving an animated talk at UCSD in the late 60's mincing around in bizarre bright green pants like a leprechaun. Maybe it was Saint Patrick's Day?

"Here are the excerpts:"

[Bishop:] *"The establishment of the Power saved the Earth from destruction by the forces of regression."*
...
[Dunn:] *"Bullshit. The Power IS the force of regression."*

Dunn is like John Savage in Huxley's "Brave New World." This is all in the tradition of H.G. Wells.

[Bishop:] *"The Power is all the knowledge mankind has gathered in its lifetime, used wisely and carefully for betterment."*

[Dunn:] *"The Power controls knowledge, keeps it under lock and key. It's stopped the gathering of any new knowledge and totally destroyed the scientific effort ... It lives off the wisdom of the past, has a stranglehold on the present, and by killing science, it kills the future."*

This reminds me of people we know who run the Cornell e-print archive and sci.physics.research. Now there is another synchronicity suggesting that Dennis Schmidt was receiving messages from the future back 25 years ago – big time. As we see in Carlo Suares's "Propositions" above, we all, deep down at the inner core of the Soul, yearn to escape, to metamorphose into something greater, to break the chains that bind us to this limited world to what the Pharisees of Orthodox Physics, now itself wildly over-speculative in desperation over a paucity of really good intuitive ideas and in denial of inconvenient "fringe" facts, call the "boring universe."

[Bishop:] *"We have enough knowledge ... the Power protects us from ourselves. It gives the people that part of the knowledge they need to make their lives better."*

[Dunn:] *"What if further research would uncover a new theory, one that might make vast areas of current theory obsolete?"*

[Bishop:] *"Impossible ... Unthinkable. Such a thing cannot happen."*

[Dunn:] *"Do you realize ... that if Sarfatti and Aspect had blindly accepted Einstein's dictum that the speed of light was the limiting velocity in the universe, they never would have conceived or proven superluminal connectedness, and we wouldn't even be here? Tachyons never turned up, spinning black holes gave random destinations with no way back, and sub-light velocities simply took too long. Only the Sarfatti-Aspect effect gave us the key to the stars ... the Power owns and controls science now. But it's a dead, useless thing you own. ... Only freedom can revive it."*

[Bishop:] *."... what was ... this vaunted science you cherish so much, based on? Human reason! ... As if the conscious, rational mind was the major driving force in human history! ... ... Even within science itself there was a protest against the iron rule of reason. Quantum theory punched the standard view of reality full of holes. Men of foresight and wisdom like Heisenberg, Bohm, and Finkelstein tried to make their comrades see the error of their ways. But it was too late ... Ruin came down on all alike ... If the Power had not come along when it did ... the destruction would have been complete. ... The Power ... is the simple faith the people have in us that provides the strength we need to control science and transform it ... into the gentle and beneficent Knowledge. It is the faith mankind gives us that allows us to give them peace and security and happiness in return."*

Wow! I had never read that far. Obviously the "Power" was Werner Erhard a kind of American Demagogue Cult Leader now largely forgotten. [xliv]

On the faster-than-light issue, we have in Wheeler's sense:

Faster-than-light without faster-than-light

On the level of warp drive the material motion is locally slower than light, i.e. on a free-float time-like geodesic inside the local light cone at each point on the ship's world line, whilst globally the ship effectively travels faster-than-light because of the inhomogeneous space-time warp between bow and stern as shown by Alcubierre.

"It is shown how, within the framework of general relativity and without the introduction of wormholes, it is possible to modify a space-time in a way that allows a spaceship to travel with an arbitrarily large speed. By a purely local expansion of space-time behind the spaceship and an opposite contraction in front of it, motion faster than the speed of light as seen by observers outside the disturbed region is possible. The resulting distortion is reminiscent of the ``warp drive" of science fiction. However, just as it happens with wormholes, exotic matter will be needed in order to generate a distortion of space-time like the one discussed here."[xlv]

The missing piece of this puzzle not known to Alcubierre at the time he wrote this paper is the fact of "dark energy." The problem is then how to "bottle" it. Mike Turner has written in April 2003 Physics Today that he does not think it is possible to "bottle dark energy." The UFO facts suggest otherwise. This issue of material warp drive is at the IT level in the sense of the IT/BIT dualism in Bohm's quantum ontology where I use Wheeler's "IT" instead" of "hidden" or "extra" variable and "BIT" for the pilot wave of both "active" and "inactive" non-classical "qubit" information. The issue of "signal locality" vs "telepathic signal nonlocality" is at the BIT level. The presponse experiments of Dick Bierman and others show the fact of "signal nonlocality" in violation of micro-quantum theory at the conscious mind-matter level.

"But there is great interest on this forum [q-mind] in the Bierman presponse experiments. Whether or not the claims will be borne out, the experiments do have enough of the trappings of good science to be not dismissable out of hand.

If the presponse effects of the kind reported by Bierman turn out to be a real feature of nature, then, as Stan says, we are at the dawn of a new age in science.

There is, then, a question of how the claimed effects could be reconciled with contemporary orthodox science: What is the simplest and most parsimonious modification of orthodox theory that could accommodate them? Must the orthodox ideas of Copenhagen and von Neumann/Wigner be rejected in favor of a radically different foundation?

I shall describe here how the addition of small non-hermitian contributions to the Hamiltonian of the brain of the conscious subject can lead to presponse effects of the kind Bierman reports." Henry P. Stapp[xlvi]

Orthodox quantum computer theory does not usually allow these "non-hermitian contributions to the Hamiltonian" that lead to non-unitary irreversible time evolution except in the case of some models of decaying quantum states in, for example, the optical model of scattering in nuclear physics. The orthodox physics literature is actually inconsistent on this issue.[xlvii] Stapp was severely rebuked in Physics Today for daring to publish this sort of work and indeed, the editors of Physical Review, to their everlasting shame, adopted a censorship policy against "fringe" topics because of the backlash against Stapp's provocative paper. The mainstream physics academic community has a hardening of the arteries and an increasing intolerance to new ideas whilst embracing even more over-speculative ideas like string theory and loop theory in a very puzzling double standard. Nobel Prize physicist Brian Josephson from Cambridge University has led the charge against this intolerance now for several years.

"It is argued that immense physical resources - for nonlocal communication, espionage, and exponentially-fast computation - are hidden from us by quantum noise, and that this noise is not fundamental but merely a property of an equilibrium state in which the universe happens to be at the present time. It is suggested that 'non-quantum' or nonequilibrium matter might exist today in the form of relic particles from the early universe. We describe how such matter could be detected and put to practical use. Nonequilibrium matter could be used to send instantaneous signals, to violate the uncertainty principle, to distinguish non-orthogonal quantum states without disturbing them, to eavesdrop on quantum key distribution, and to outpace quantum computation (solving NP-complete problems in polynomial time)." Antony Valentini[xlviii]

Valentini may be wrong about relic particles? We don't need them. Exotic vacua may do the trick?

Tony Smith then wrote on Feb 6, 2004:

"A review of Satori, Kensho Novel #3 of 4,
by Dennis Schmidt (Berkeley 1981).
Unless otherwise noted, quotes are from the book.

On a web page[xlix] the Wanderling says:"

"Satori is the spiritual goal of Zen Buddhism. (in Chinese: wu.) Satori roughly translates into individual Enligthtenment, or a flash of sudden awareness. ... A brief experience of Enlightenment is sometimes called Kensho. ... In describing the Enlightenment of the Buddha and the patriarchs, however, it is customary to use the word Satori rather than Kensho, the term Satori implying a deeper experience. The feeling of Satori is that of infinite space."

"Ben Holmes has written a review of the book Satori. I quote the back cover."[l]

"Satori - The Earthmen are coming... For a thousand years the men and women of Kensho have been evolving a way of living, a civilization that combines technology and spirit in a fashion that the visionaries of old Earth could hardly imagine as the two essential parts of a healthy whole, rather than as deadly enemies out to destroy each other. In another century Kensho would have become invulnerable to the violence of

Earth, but the starship bristling with fearful weapons ... is orbiting the planet now. Is there no alternative but to submit to the conquerors? Must the New Humanity return to the sad old ways of Earth? There is one possible way out - but if it fails, the people of Kensho will pay the price of annihilation."

"Earth is controlled by an est-like Cult of the Power. The Bishop of the Power is interrogating a human named Dunn on a starship from Earth that is threatening the humans of the planet Kensho as shown in the excerpts from the book above.

The humans of Kensho must compete against this Earth-based spaceship that has not only weapons that can incinerate a planet, but also machines used by the Power for mind-control that can gather all the data in a mind and then record, correlate, and analyze all that data and then imprint on that mind new data consistent with the Power.

However, the humans of Kensho have developed mental structures, "groups of neurons attached to each other in what seemed to be a closed feedback loop", that allow them to live in symbiosis with Mind Brothers, each of which is a 3-space/4-space-time projection of a 'single creature... from another, higher dimension."

The same idea of a hyperspace intelligence is in Jacques Vallee's "Fastwalker" and in Carlo Suares' "Eight Propositions." This is where my physics model of consciousness using telepathic "signal nonlocality" comes in. Signal nonlocality violates microquantum theory which is only a limiting case of a more general post-quantum theory. Martin Gardner explained an early version of my theory in 1975 in "Magic and Paraphysics" that Dennis Schmidt may have read?

"Kensho symbiotic humans see the Earth Power mind-machines as 'Primitive ... the mere gathering of large quantities of data hoping some pattern will automatically emerge ... a crude version of empiricism ... At first blush it seems to make sense. But ... collections of data are just that, collections of data. Until they're put into some kind of meaningful order, they're useless. So naive empiricism was replaced by the techniques of developing a hypothesis first, and then looking for the specific facts that prove or disprove the conjecture' ... With their mental symbiosis, Kensho humans could survive where the Earth Power men and mind-machines could never reach: 'The abyss, the void ... where solidity disappeared, time halted, and ... the self splintered into ... fragments. The world, the universe, simply evaporated like a wisp of morning fog.'"

From Fastwalker to Skinwalker

"Memo to file: if you're one of those high-strung people prone to be rattled by the occasional bulletproof wolf, flying refrigerator, disappearing/reappearing interdimensional gateway, lumbering giant humanoid, dog-incinerating luminous orb, teleporting bull, and bloodlessly eviscerated cow, don't buy a ranch, even if it's a terrific bargain, whose very mention makes American Indians in the neighbourhood go "woo-woo" and slowly back away from you. That's what Terry Sherman ("Tom Gorman" in this book) and family did in 1994, walking into, if you believe their story, a seething nexus of the paranormal so weird and intense that Chris Carter could have saved a fortune by turning the X-Files into a reality show about their life. The Shermans found that living with things which don't just go bump in the night but also slaughter their prize livestock and working dogs so disturbing they jumped at the opportunity to unload the place in 1996, when the National Institute for Discovery Science (NIDS), a private foundation investigating the paranormal funded by real estate tycoon and inflatable space station entrepreneur Robert Bigelow offered to buy them out in order to establish a systematic on-site investigation of the phenomena." John Walker, the complete review is at http://www.fourmilab.ch/fourmilog/archives/2006-02/000658.html

To Robert Bigelow, for his brilliant and farsighted vision.

To the Gorman family, for their extraordinary courage.

To JACK SARFATTI,

A BRILLIANT MAN WHO SOMETIMES DRIVES people CRAZY.

I HOPE YOU LIKE THE BOOK BUT LOOK FORWARD TO HEARING YOUR THOUGHTS NO MATTER WHAT.

George Knapp
DEC. 2005

On Mar 9, 2006, at 2:22 AM, George Knapp wrote:

"Jack, Thanks for forwarding the review of Skinwalker. I don't know anything about your friend John Walker, but I thoroughly enjoyed his comments and got a good chuckle from several of his well-crafted lines. Clever stuff, especially the 1997 article. Damn, that's a fine if twisted take on the phenomena. If humans are rattled by the idea of alien spacecraft zipping around in our skies, imagine what the reaction will be if it can be proven that all of these silly UFOs are nothing more than ETs copulating above our heads. ... In a sense, Mr. Walker is saying that Earth is the galactic equivalent of Las Vegas. Horny space swingers and conventioneers travel here with their secretaries or mistresses because they know that what happens in the atmosphere stays in the atmosphere, and the little woman back home in the Crab Nebula doesn't need to know. Walker's more serious criticisms are more than fair and I understand why he posits them. That doesn't mean I agree with them. So, in the spirit with which they were offered, I'd like to respond. 1) When he compares the NIDS trailer to the VW van driven by the Lone Gunmen, he is way off base. A better comparison would have been Ken Kesey's Magic Bus. 2) The statement that most of the interesting stuff ended as soon as NIDS arrived is incorrect. The book admits that the level of activity changed once the team arrived and started "hunting" the for the source. The NIDS team and its scientific board came to believe that the "entity" on the ranch was smart, elusive, and purposeful, displaying a kind of gamesmanship similar to "trickster" tactics that have been reported elsewhere. It didn't like being stalked and it rarely revealed itself in the same way or same spot, at least, that's the impression of the team, the ranchers, the Utes, the neighbors, and others. If readers aren't willing to at least entertain the possibility that this is an accurate description of how things unfolded or that an unknown sentient intelligence could accomplish such things, then I don't know why such readers would even finish the book. While writing the book, Colm and I were very aware that readers would probably have to engage in a willing suspension of disbelief, to borrow a film term, just to absorb the ridiculous array of events that occurred. ... We told the story as accurately as we could. .... The scientists and staff personally witnessed a hell of a lot, not only random lights, orbs, and structured craft, but plenty more, including the tunnel of light that appeared from which a large creature emerged, the dinosaur creature that was shot out of a tree, the mind-meld incident when a black cloud engaged a physicist, dramatic and inexplicable reactions by humans, animals, and sensors, and a lot more. The team didn't personally witness the mutilation of livestock as it occurred, but they were on hand to examine and analyze the physical evidence. Ditto for the incident with the bulls. Ditto for the vandalism of the video equipment by a force that was invisible to adjacent cameras. Ditto for the strange ice circle that formed on a pond. They endured a hell of a lot of weird stuff, maybe not as much as the ranch family endured, but more than most of us will ever see in a lifetime. ... Journalism isn't a hobby for me. This is my job. I don't make stuff up and I don't put my name on a written account that isn't truthful. I've been reading the field reports about the ranch for a lot of years now. The reports weren't written so that they would eventually be compiled into a book. No one at NIDS wanted this stuff to be released, not because of any hidden agenda, but because of the knowledge that such a collection of strange info might harm the reputation of the

organization. It took a long time to talk the principals into allowing me to write something. ... □Regards to all,□ □GK"□

I met Jacques Vallee in 1975 or so. He was on the periphery of the Physics Consciousness Research Group working with Ira Einhorn[li] and came to several of our gatherings. Saul-Paul Sirag and I took him to Francis Ford Coppola's house in 1976 with Uri Geller. That's how Jacques connected with Stephen Spielberg and became the technical consultant to "Close Encounters of the Third Kind." The character Francois Truffaut plays in the film is based on Vallee. Truffaut was working with another late friend of mine Jean Reisser Nadal. Some of that story is in "Destiny Matrix." Vallee's sci-fi novel "Fastwalker" with Mel Torme's son, Tracy (of TV sci-fi "Sliders" on parallel universes) is based on "fact" very much like Esalen founder, Michael Murphy's novel about the Russian Psi Movement in the Cold War "An End to Ordinary History" is not really pure fiction. I lived in the apartment on Telegraph Hill on 2 Whiting Place that Michael describes as Jacob Atabet's abode. "Atabet" is obvious code for Qabalistic "Aleph-Bayt" alphabet soup. (Grin). Michael lived there before me and I got the place from him. Vallee and Torme describe the same kinds of phenomena in their novel that Eric Davis lists in his MUFON 2001 paper. Vallee today is a venture capitalist in Silicon Valley[lii] who publicly claims he no longer does UFO research yet he, until recently at least, advised Las Vegas Real Estate Tycoon Robert Bigelow on the subject in Bigelow's NIDS project [liii]where Eric Davis had been working when he wrote the MUFON 2001 paper. Vallee continues to attend select meetings on UFOs and has been less than candid here. Here are some tantalizing snippets from the "novel" with my comments. Thanks to Robert Jones of the Caffe Trieste Irregulars who did the background research for me here on Vallee's essentially factual report thinly disguised as *"science-fiction."*

MK-ULTRA at Cornell?

"MK-ULTRA had been approved by Allen Dulles ... on April 13, 1953..." P. 50

"They signed up Cornell." P. 50

I arrived at Cornell at age 17 on full scholarship arranged by Walter Breen who was working on a gifted children project under Professor William Sheldon of Columbia University Psychology Department financed by Eugene Mc Dermott, a co-founder of Texas Instruments, with strong ties to US Intelligence from WWII. Phil Morrison, one of the founders of SETI was my advisor.

"As early as April 9, 1952 you will find a memo that suggested looking at the influence of drugs on human paranormal abilities." P. 52

Puharich in the Army working on psychedelics in early 50's

Dr. Andrija Puharich did that work for the Army at that time. I later met Puharich in 1974 during my involvement with Uri Geller as told by Martin Gardner in "Magic and Paraphysics", in "The Star Gate Conspiracy" by Picknett and Prince and in "Destiny

Matrix." Walter Breen attempted to induce paranormal powers in us kids (1954 – 56), although no drugs were used. Agents allegedly from New Mexico who I met monitored his project. I also met CIA Chief of Station Harold Chipman in mid 80's who was very much part of this project. Chipman wrote some of the episodes in the TV series "The Enforcer"based on his CIA career. He also wrote a treatment for a screen play "The Union" about a rogue CIA group like Vallee's "Alintel" using remote viewing against the Soviets. He was an important part of MK ULTRA going back to the fifties and had a hand in the SRI remote-viewing (RV) project of Puthoff and Targ with astronaut Edgar Mitchell and Brendan O Regan. Police Commissioner Pat Price was a close associate of "Chip's" and Price was a "star" of the RV project along with Uri Geller and Ingo Swann. Puthoff was not aware of Chipman's role in what he was doing at the time.

Alien Intelligence

"Alintel ... Alien Intelligence ... Who would form a unit like that? ... A president ... designed Alintel as a black project: No records, no traces, no bureaucratic trails ... The UFO phenomenon has turned out to be far more complex than anyone had dreamed."

"So you believe it's real?"

"Oh we knew that from the start ..." P. 86

"We're a rogue element ... Newcomers are being brought in- military men, scientific types. It's a shift in long-standing policy ... The phone rang at the CIA, in the office of James O' Grady, Ph.D., a thirty-five-year-old scientist who had been assigned the task of looking after all the bizarre, undocumented and unwanted data ... he was assigned the job of 'Keeper of the Weird'..." Pp.89 – 90

"How did your big stick impress the Soviets ...?"

"Alintel showed them how the UFO material could be a source of immense energy, and could make the balance of power go tilt." P. 92

Indeed, the exotic vacuum stress-dark zero point energy/matter density Einstein tensor field t_{uv}(ZPF) for practical UFO metric engineering of free float "acceleration field" warp drive and Star Gate construction is from the two complementary models of quantum gravity, i.e. Loop and String theories:

t_{uv}(ZPF) ~ (Witten String Tension)(Loop Quantum of Area)$^{-1}$[(Loop Quantum of Area)$^{3/2}$(Intensity of Local Vacuum Coherence) - 1]g_{uv}

The Einstein field equation of General Relativity to a good approximation is then

$$t_{uv}(\text{space-time warp}) + t_{uv}(\text{ZPF}) \sim 0$$

where

$$t_{uv}(\text{space-time warp}) = (\text{Witten String Tension})(R_{uv} - (1/2)Rg_{uv})$$

and a few other subsidiary equations of course. The UFO physics problem of the "acceleration field" is solved and demystified at the fundamental conceptual level.

Gorby caves in?

"We think the Kremlin shift in policy towards America had something to do with the captured UFO hardware..." P.93

"He recited the Mission's purpose ... To enhance the credibility of the extraterrestrial threat ... to discredit those who might guess that we do not have ultimate control over the UFO situation; and to flush out any competition that might interfere with our action from within the Intelligence community. ... Academia is a control system, too, but it's a closed one, and the rules are extremely strict. Their margin is nil. ... The UFO phenomenon must be a part of Nature ... Yes, but it is controlled by a non-human form of consciousness. Ergo, the study of it doesn't belong in science at all, it belongs in Intelligence. Meaning counter-espionage" P. 97

"The Air Force had initiated a massive study. A huge budget for the time. It was entrusted to the Battelle Memorial Institute. Then in 1953 ... the CIA organized a panel." P. 109

"We were a small part of a very big project that went all the way back to the fifties, when the military was collecting and analyzing foreign technology data." P. 131.

Yes indeed, you can say that again. Vallee's opinion on UFO Cults is given on P. 134. Note that Vallee's book "The Network Revolution" documents the penetration of Werner Erhard's est into Silicon Valley in the late seventies and early eighties. Werner Erhard also threw a lot of money at big physicists in the late seventies. Leonard Susskind mentions that:

"I met Hawking and Gerard 't Hooft in the attic of Werner Erhard's house in San Francisco. Erhard was a fan of Sidney Coleman. Dick Feynman, myself, and David Finkelstein were his gurus. And of course we didn't give a damn about his silly business, but we loved his cigars, we loved his liquor, we loved the food that we got from him, and he was fun. He was very, very smart." [liv]

"What do we care if we understand the physics or not? ... It doesn't really matter how these things fly."

"It does matter to us, as scientists .... It's not enough to know that they are real." P. 143

Alluding to Dan Smith's, Laurance Rockefeller's and Robert Bigelow's funding of UFO projects:

"What matters is that influential people ... folks with mansions and private jet helicopters ... and billion dollar stock portfolios – now realize that the phenomenon ... impinges on their business on their ability to manage the planet ..." P. 144

"What the believers perceive as the Eschaton is really a planetary transition. That's what we need to manage. ... We're making a transition from being monkey-people to something else. That's the real message. We're managing an Apocalypse." P.147
See the Dan Smith interview in Paranoia Magazine.[lv]

"We find out what scared the shit out of Gorby's boys." P.167

The Gorbachev Foundation in San Francisco was run by people with close ties to New Age UFO people and money in the period 1999-2000. One example of that is Danny Sheehan who was John Mack's attorney in his tenure battle with Harvard over Mack's UFO abduction research. Sheehan organized the Gorbachev Foundation State of the World Forum. I attended key planning meetings with Sheehan as well as the conference.

The Star Gate Conspiracy?

"The New Age is the religion of the future ... We're the New Pharaohs." P. 168

"We're dealing with nuts and bolts craft, weighted at over a hundred tons." P.173

"Tell them to rerun the Doppler tests." P. 173

The exotic vacuum zero point "dark" energy/matter stress density tensor weightless warp drive will show a reverse Doppler effect, i.e. universal redshift as UFO approaches from the attractive dark matter layer at the bow and a universal blue shift as the UFO recedes from the repulsive dark energy layer at the stern. This is opposite to the normal motional Doppler frequency shifts.

"It is shown how, within the framework of general relativity and without the introduction of wormholes, it is possible to modify a space-time in a way that allows a spaceship to travel with an arbitrarily large speed. By a purely local expansion of space-time behind the spaceship and an opposite contraction in front of it, motion faster than the speed of light as seen by observers outside the disturbed region is possible. The resulting distortion is reminiscent of the 'warp drive' of science fiction. However, just as it happens with wormholes, exotic matter will be needed in order to generate a distortion of space-time like the one discussed here." [lvi]

Reverse Doppler Effect from the Flying Saucers

Bruce Cornet had reported such a "reverse Doppler effect" from the sound coming from flying saucers without knowing of the theory I propose here.[lvii] I am not corroborating his allegation here simply mentioning that it exists.

Alien Raj origin of all religions?

On the UFO Contact origin of all major world religions see P. 178.

Abraham
Moses
Saul of Tarsus
Mohammed
Joan of Arc
Joseph Smith (Mormons)

"The craft had no plasma thrusters ... "
"... luminosity profiles?"
"They don't match the analysis of the negative."
"... surface reflectivity?"
"... it resembles a constantly moving film with an approximate optical depth of five microns, rather than a fixed metallic skin."
"... the membrane seems to be agitated by random interface phenomena at the molecular level."

"What's it doing now ...?"
"It generates a very weak magnetic field, but it doesn't emit any light any more."
Pp. 178-79

"To their right was the most magnificent sight of all: a grounded saucer a hundred feet in diameter, silvery-gray in color, with an occasional glint of gold. The sound that came from it was a high-pitch melody that followed none of the known laws of music. The craft's surface shimmered under the spotlights, as if it were spinning rapidly."

Strong gravimagnetic field $\mathbf{H} = (g_{01}, g_{02}, g_{03})$, but with with $G^* \gg G(\text{Newton})$. See Wheeler and Ciuofolini "Gravitation and Inertia." Ray Chiao uses this field in "gravity radio", but at much weaker G(Newton) level.

"The disk had a superstructure like a cupola on its upper shell, with pulsing colors that resolved into successive levels of blue, red and green. The UFO was softly humming .... Its physical shape seemed to change ever so slightly from minute to minute defying perspective, inducing a mild sense of vertigo..." P.180-1

Strong short-range zero point vacuum fluctuation induced gravity lensing? The sound frequency-shift z pattern would mimic the light color-shift strong (anti) gravity z-pattern as well.

GM(source of warp field)/c² is replaced by $-4\pi\Lambda_{zpf}r^2dr$

For a thin spherical shell of exotic vacuum boundary layer of radius r and thickness Δr

The *metric warp factor* in this over-simplified, to be sure, static spherically symmetric toy model is then $1 + 8\pi\Lambda_{zpf}r^2\Delta r / r'$ at a field point $r' > r$ outside the "shell", i.e. fuselage of the ET alien flying saucer like in the underground lab in the film "Independence Day". The physically measured time differential at the distant field point r', at the detector, is

$$d\tau = \left(1 + \frac{8\pi\Lambda_{zpf}r^2\Delta r}{r'}\right)dt$$

The universal color shift z between r and r' is computed from

$$dt(r) = dt(r')$$
$$r(\text{source}) < r'(\text{detector})$$

Therefore

$$\frac{d\tau}{\left(1+\dfrac{8\pi\Lambda_{zpf}r^2\Delta r}{r}\right)} = \frac{d\tau'}{\left(1+\dfrac{8\pi\Lambda_{zpf}r^2\Delta r}{r'}\right)}$$

$$\frac{v'}{\left(1+\dfrac{8\pi\Lambda_{zpf}r^2\Delta r}{r}\right)} = \frac{v}{\left(1+\dfrac{8\pi\Lambda_{zpf}r^2\Delta r}{r'}\right)}$$

$$\frac{\Delta v}{v} \equiv \frac{v'-v}{v} = \frac{v'(\text{detector})}{v(\text{source})} - 1 = \frac{\left(1+\dfrac{8\pi\Lambda_{zpf}r^2\Delta r}{r}\right)}{\left(1+\dfrac{8\pi\Lambda_{zpf}r^2\Delta r}{r'}\right)} - 1$$

Consequently the weird UFO color shifts are explained by the strong warping of the fabric of space-time by controlled Josephson "weak link" vacuum ODLRO phase interference manipulation of the local zero point energy density where

$$\Lambda_{zpf}(dark - energy) > 0 \rightarrow \frac{\Delta v}{v} > 0\,(UFO \,/\, blue - shift)$$

$$\Lambda_{zpf}(dark - matter) < 0 \rightarrow \frac{\Delta v}{v} < 0\,(UFO \,/\, red - shift)$$

Note that astronomers use

$$z \equiv \frac{\lambda' - \lambda}{\lambda} = \frac{\frac{1}{v'} - \frac{1}{v}}{\frac{1}{v}} = \frac{v}{v'} - 1 = \frac{\left(1 + \frac{8\pi\Lambda_{zpf}r^2\Delta r}{r'}\right)}{\left(1 + \frac{8\pi\Lambda_{zpf}r^2\Delta r}{r}\right)} - 1$$

$z < 0 (blue - shift)$

$z > 0 (red - shift)$

Bruce Cornet claims to have measured this effect from real flying saucers using sound waves. Since this is a warp effect, all kinds of waves will be affected in the same way. Cornet worked as scientific director of NIDS in Las Vegas funded by Hotel Mogul Robert Bigelow who also owns Bigelow Aerospace and is the second Howard Hughes. Jacques Vallee, Hal Puthoff, Colonel John Alexander (USA-ret) and Dr. Christopher Green MD (CIA-ret) were all on the Scientific Advisory Board of NIDS where Eric Davis also worked before moving to Hal Puthoff's IAS Austin after writing the "teleportation" paper for Franklin Mead of USAF advanced propulsion group at Edwards Air Force Base. Vallee's book is not really sci-fi, it is what Tim Leary called "science-faction", i.e. allegedly thinly-disguised top-secret Black Operations. Most of the staff at NIDS before it went underground, officially folded, were former FBI, police, CIA, Army Special Forces etc. The book "Hunt for the Skinwalker" by Colm Kelleher and George Knapp that came out at the end of 2005 continue this story into the uncanny Twilight Zone of a malevolent telepathic alien intelligence on Bigelow's Utah Ranch known to Native Americans. There are at least two other locations where the phenomenon has happened also associated with American Indian sacred grounds.

The radial warp distortion is

$$dR = \frac{dr}{\left(1 + \frac{8\pi\Lambda_{zpf}r^2\Delta r}{r}\right)}$$

For a dark energy shell this is a contraction of the radial distance relative to the transverse distances because the radially-oriented rods expand relative to the transverse rods. For the dark matter shell it's a radial expansion as usual because the radial rods shrink relative to the transverse rods. There is no longitude-latitude distortion here as there is in the Puthoff metric engineering model.[lviii] When the shell of exotic vacuum is spinning we need the Kerr vacuum solution with the gravimagnetic field and the math is much more complicated than the toy model above. Puthoff has never been able to solve that problem in his PV model as far as I know. It's solved in Einstein's GR.

"The Fastwalker had changed shape again, and it began to glow with the intensity of a blaze. What they saw now was a rounded mass of light ninety feet wide and forty feet high. Its intensity resolved itself into successive layers of bluish and reddish radiation."
P. 181

Big Bang?

"The power in this room is virtually unlimited."
"No wonder the Reds fell to their knees." P. 183

"We will now measure the Doppler effect ... 1,801 km/h it read ... that baby is flying away from us at over a thousand miles an hour right now."

"But it's not moving at all." P. 184

This, obviously, is not a motional Doppler shift at all, but my gravity shift as in my equations above. Vallee, of course, had no idea of this physics when he wrote the above. However, I thank him for his empirical descriptions like Eric Davis's in MUFON 2001. Explaining this stuff is a piece of cake once you have the correct ideas.

"We must be inside the Fastwalker ..."

"How can the inside be so much bigger than the outside?" P. 198

Elementary my Dear Watson, simply Einstein's equation X above with the exotic vacuum zero point dark energy source controlled by the local vacuum coherence intensity via a Josephson weak-link.

"... trying to keep her balance on the tilting, undulating ground. Somehow she stayed on her feet."

To be expected in a dynamic space warp including staying on your feet and not falling down as you would in an ordinary earthquake. Paul Hill essentially explains this in his discussion of the "acceleration field." [lix]

Masters of Hyperspace

"We are from the Multiverse ..." P.199

"It will be years ... before humans achieve contact with the consciousness of space." P.200

Time's up, Bhubba. Since space-time emerges from the phase ripples in the local giant quantum vacuum wave and since that wave admits signal nonlocality, that wave is the conscious "Mind of God" in Hawking's sense in my humble opinion. Hence space is conscious. Our mind fields piloting our brain function are also local giant quantum waves in the image of the Mind of God sort of like The Bible suggests. That's my breaking of the cipher of the Da Vinci Code, the Meaning of The Grail. Take it or leave it.

"The Web incarnates the dark intuition that Henry Adams, looking into a future ruled by the dynamo, articulated almost a century ago: that we do not inhabit a universe at all, but a Multiverse." Erik Davis, Techgnosis p. 244

"The G-Engines Are Coming!"

1956 article by Michael Gladych cited on p.3 of Nick Cook's
"The Hunt for Zero Point"

They really are coming in my opinion. Or, rather, they are allegedly already here. That's what the flying saucers are, if they real and not an elaborate hoax. My physics research suggests more and more and faster and faster as time goes on that they are not a hoax. All scientific rational knowledge about the real world, however, is uncertain (Bayesean probability) and, in principle, must be falsifiable as explained by Sir Karl Popper and applied in the financial markets by his student George Soros. Religious faith in extreme form is simplistic and dangerous as we see in the current war on terror and throughout history.

" 'All matter within the ship would be influenced by the ship's gravitation only,' Lear apparently said of the wondrous G-craft. 'This way, no matter how fast you accelerated or changed course, your body would not feel it any more than it now feels the tremendous speed and acceleration of the Earth.' The G-ship, Gladych explained, could take off like a cannon shell, come to a stop with equal abruptness, and the passengers wouldn't even need seat belts." Nick Cook Pp. 3-4.

This is an amazing statement considering it was made in 1956, although Paul Hill had deduced such things purely empirically from the flying saucer evidence. It is in complete accord with my own independent theoretical physics research. Scientists are not supposed to reject evidence. The prime-time mass media millionaire physicists on John Brockman's client list[lx] scoff at the kind of hard evidence that Bruce Maccabee, a US Navy physicist and Eric Davis another physicist present below. Little do the mainstream pundits realize that the discoveries of the new cosmology of dark energy since 1998 are consistent with the UFO physical evidence. They would rather sell the public unsubstantiated speculations of "strings" as "vibrations of pure energy" in 10-dimensional space and of radiating black hole. Both of these examples may well prove true in the end, but there is no evidence that they are. There is today much more evidence for UFOs performing seemingly impossible maneuvers than there is from strings and radiating black holes.

Paul Hill's Acceleration Field Propulsion

"The names force fields and acceleration fields are interchangeable. Here we probably have pay dirt. Fortunately so, for we have almost eliminated all of the other possibilities by virtue of their being inconsistent with observed facts. In contrast, force field propulsion will be found to be consistent with all reported observations that fit a general UFO pattern, no matter which type of UFO ... One of the general observations of UFOs is the lack of visible external components easily recognizable as part of the propulsion system. Force field equipment could be located within the shell of the UFO, which the field would penetrate nondestructively while the field generators remain hidden within."

Paul Hill is not quite precise enough here. The difference is the equivalence principle. The true "acceleration warp field" is universal independent of the mass that is responding to the field. That is, all parts of the object approximately accelerate the same way to the extent that the tidal curvature field is slowly changing over the size of the whole object. This is usually a very good approximation to make. In contrast, other force fields like the electric field, for example to not make all objects accelerate the same way. The response of the object to the electric field depends on the variable ratio of the electric charge to the mass. The old Galilean equivalence principle says that the inertial mass in Newton's second law

$$\mathbf{F} = m\mathbf{a}$$

is identical to the gravitational mass in Newton's static inverse square law of gravity

$$\mathbf{F} = -Gm M\mathbf{r}/r^3.$$

Therefore,

$$\mathbf{a} = -GM\mathbf{r}/r^3$$

The mass m of the test particle responding to the acceleration warp field cancels out of the problem. That's what you need for a true warp drive. It does not happen with a hypothetical electromagnetic stress drive like Jim Corum tried to do at SARA and ISR. Bold face means 3-vectors, i.e. arrows in space with both magnitude and direction that add according the law of parallelograms if you can ignore curvature. An electromagnetic stress drive is just another impulse propulsion method with those damn g-forces that we do not want! We do not want to hold on two our hats and have our stomachs go to the floor or upchuck our lunch when doing a 180 degree turn at Mach 10 in a dogfight with some nasty alien from the Universe Next Door.

In the case where the source mass M is not a real mass but a ball of radius R of exotic vacuum of uniform zero point energy density replace GM by $4\pi c^2 \Lambda_{zpf} R^3/3$. This allows a repulsive anti-gravity acceleration warp field because Λ_{zpf} can be of either sign.

Einstein's general theory of relativity adds a richer conceptual structure to this early form of the principle. The role of quantum modifications to the equivalence principle is still not completely settled theoretically and experimentally. This gets us to quantum gravity and the attempts of string theory and the loop theory.

See pages 95-97 of Paul Hill's book for more sober qualitative theoretical speculations. Section VII Pp. 98 – 108 has case studies consistent with Bruce Maccabee's suggesting that Phil Klass's and Carl Sagan's testimonies on NOVA (cited elsewhere in this book) are suspect as misinformation and disinformation, whether wittingly or unwittingly, without speculating on the motives of either of these gentlemen, all undoubtedly "honorable men" (Shakespeare's "Julius Caesar") indeed "No possible doubt whatever." (Grand Inquisitor, Gilbert and Sullivan's "Gondoliers"). See also VIII, which concludes:

"UFOs are propelled by a force akin to gravity, but of an opposite nature." P. 118.

Paul Hill's speculations on "anti-gravitons" should be discarded. The physics is now understood and is shown in this book. It was not at all understood by anyone human back then when he was writing. In the simplest case from Einstein's general relativity and quantum theory the relevant equation in the simplest static weak field limit is the exotic vacuum zero point energy density Poisson gravity local field equation

Laplacian of the exotic vacuum universal gravity potential energy per unit point test particle[lxi] is proportional to $c^2 \Lambda_{zpf}$ that, in turn, is determined by the local vacuum coherence field as shown in detail in this book.

UFO Acceleration Fields: The Evidence

Bruce S. Maccabee, Ph.D.

ABSTRACT

Visual and photographic sightings of UFOs carrying out "impossible" high-speed maneuvers are presented for study. For the first time we are able to quantify the amazing acceleration of these craft.

Prologue

Herbert C., Army private, was stationed at Fort Sill, near Lawton, Oklahoma, in June, 1968. He and another man were standing guard at about 4:30 a.m. near the airport when he noticed a lighted object, which appeared to be over the mountains about 10 miles away. He thought it was an airplane. He looked away from it for a few seconds and when he looked back where the object had been, it wasn't there. Instead, it was flying past the control tower and was within a few hundred feet of him. It had traveled perhaps as much as ten miles in a few seconds. He could see that it was a glowing sphere, perhaps (30 feet) in diameter. It had made no noise so he pointed it out to the other guard. They watched the yellow orb, like a small, hazy sun, as it moved around and lit up ground structures causing obvious shadows. After perhaps 10 - 20 seconds it suddenly rose upward at about a 45° angle for a short distance, made a 90 degree turn and then zipped off "faster than he has seen missiles in the Army."

Lillian Sargent, housewife, was standing in the back yard of her Greenfield, Mass. home in June or July of 1947 (exact date unknown). She was about half a mile west of a steeply rising small mountain that forms the eastern border of Greenfield (home of the "Poet's Seat" tower). Looking to the east, she happened to notice two round, metallic-looking objects suddenly come into view over the mountain, flying westward. From her location on Maple Street, the top of the mountain was at an angular elevation of about 30°, so she was looking upward at a reasonably steep angle. Suddenly, these two objects made a right angle turn, and headed northward away from her

It was hot in the summer of 1956, but during one night no place was hotter than the Air Force bases at Bentwaters and Lakenheath in England. The ground control radar picked up several targets, which were stationary for long periods of time (minutes) and then suddenly moved away at high speeds. According to the Condon Report, 3 "As we watched, the stationary target started moving at a speed of 400 to 600 mph in a north, northeast direction "There was no slow start or buildup to this speed — it was constant from the second it started to move until it stopped." At another time radar detected an object "making sharp rectangular course changes and this maneuver was not conducted by circular path but (by turning at) right angles at speeds of 600 – 800 mph.." (emphasis added)

During the night of November 16-17, 1986, a 747 jumbo jet freighter was flying southwestward over Alaska. Piloted by a Japanese crew of three and headed for Tokyo after a stop at Anchorage, the plane was entering Alaska air space when Captain Terauchi first spotted lighted craft traveling below and to his left. He thought they were U.S. Air Force jets and paid no more attention. Minutes later the two "jets" had disappeared and two "spaceships" had suddenly appeared in front of the 747, which were traveling at about 600 mph, where the whole crew saw them. The crew reported these ships to the Anchorage Air Route Traffic Control Center and there began a series of incidents and sightings, which have made history. The Federal Aviation Administration eventually investigated the sighting and learned that these "spaceships" maintained a roughly fixed distance in front of the plane for about 10 minutes, at which time they suddenly departed to the left so fast that they seemed to disappear.

It was about 1:15 a.m. on May 1, 1988 and Ed Walters, standing on the shore of the Santa Rosa Sound with his special stereo "SRS" camera, was about to take a second stereo picture of a UFO which he had just photographed hovering over the water. As he sighted through the viewfinder he was surprised to find that it wasn't there any more. He took his eye away from the viewfinder to look for it and realized it was over his head. Suddenly everything went white...an abduction had begun. Subsequent analysis of the stereo pair of photographs showed that the object was about 450' away when he first saw it. Seconds later it was stationary over his head.

Introduction

What are these things that come and go in the night and in the day so rapidly that we have a hard time seeing them? The Shadow Knows, but he's not telling, so we'll have to figure it out by ourselves. Aside from shape and speed, the most unusual and unbelievable UFO characteristics which have been often reported are the "right angle" turns and the sudden disappearances of apparently solid objects in the clear sky. Of course, the witnesses who make these reports are not perfect observers. Perhaps the turns were not perfectly abrupt, they only appeared to be. Perhaps the objects did not disappear in the sense of material objects vanishing without a trace, perhaps they only appeared to vanish. Physics and technology, as we know it, indicates that it is physically impossible for substantial objects to make instantaneous turns or disappear while in plain view. For this reason, conventional scientists have "solved" the problem posed by these observations by rejecting the UFO reports altogether. Ufologists have accepted the reports, perhaps reluctantly, while making excuses for these seemingly impossible feats of the UFOs. Nevertheless, credible people make these reports. Many years ago (1947) Lillian Sargent, my grandmother, reported to her family that she saw two round, silvery "flying saucers," each of which made (what appeared to her to be) a right angle turn. The saucers were at a reasonably high elevation angle as they moved westward and then abruptly turned north, so she was in a good position to see the turn. Now, grandma was not an aerodynamics expert or even a mechanic, but she knew that something was not right here: "things" do not make right angle turns. Turns are curved. Nevertheless, she stuck to her story. Hence, we are left with the quandary of believing a grandmother's report while doubting the "right angle turn." (If you can't believe your grandmother, who can you believe? Rhetorical question, only.) (Footnote: this was at a time when many people were reporting sightings throughout the United States, and the reporting of sightings was considered almost like a patriotic duty. However, because of a lack of "hard evidence," the press began to turn against these sightings, and witnesses were being called crazy or deluded or hoaxers. Initially, Grandpa was happy to have his wife tell people that she saw saucers, but then when the "counter reaction" set in, he told her to shut up and never mention it again. After that she kept silent for over 20 years. I first learned of it from my mother in the late 1960s. I subsequently interviewed my grandmother.) Disappearances in the clear sky are equally enigmatic and have been reported since 1947. The first official reference to the ability of UFOs to travel at high speed, and even to disappear, is in the draft of an intelligence collection memorandum written for Brigadier General Schulgen by Lt. Col. Garrett in October 1947. (This memorandum, which was released to the public by the Air Force in 1985, was written in response to General Nathan Twining's letter of September 23, 1947, which said that flying saucers are "real and not visionary or fictitious" and outlined reasons for collecting intelligence information about flying saucers.) Lt. Col. Garrett based his statements on sightings, not by housewives, but by "many competent observers including USAF rated officers" and listed a number of "commonly reported features that are very significant." This list includes "the ability to suddenly appear without warning as if from an extremely high altitude" and "the ability to quickly disappear by high speed or by complete disintegration." (emphasis added) Sudden appearance and disappearance could be two aspects of the same capability: extremely high acceleration or deceleration. (Note: the suggestion that a flying saucer could disappear by "complete disintegration" indicates just how puzzled the top Air Force officers were by the sighting reports. For the purposes of this paper, I assume that visual "disintegration," suggesting at the very least the termination of the light reflecting or light blocking (opaque) characteristics of a solid, stationary object, does not occur. Of course, I could be wrong!) A high quality report of a disappearance event was made to Dr. James McDonald and later to the American Society of Newspaper Editors in 1967 by William Powell, a general aviation pilot, and his traveling companion, Muriel McClave. According to Mr. Powell, they were flying northward at about 4,500' near Willow Grove, Pennsylvania, on May 21, 1966 at 3:15 P.m. when they saw an unidentifiable object following some jets that had just taken off to the north from the Naval Air Station at Willow Grove. Powell at first thought it was an aircraft, and then realized that he could see no vertical tail fin. "I couldn't determine any tail on this object. And the more I kept peering at it, I sort of whimsically thought it was a flying saucer," he told the newspaper editors. "Hey Mick (Muriel), look at that flying saucer out there," he said and she immediately looked and saw it. Then they saw it make what appeared to them to be an abrupt, flat (no banking, no slewing) right turn of about 160° and head toward their plane. As it approached from the left side of the aircraft, the angular size increased and Powell tried to "envision some wires or something hanging down from it that looked like a weather balloon or elongated weather balloon, but it was exactly what I had heard and read about...so-called UFOs." The sky was clear except for some cumulus clouds above, and the visibility range was estimated at 15 miles, so they had a clear view of this object. Powell estimated that it came to within about a hundred yards of his aircraft before it passed by to the right. "It was a saucer shape with a slight raised dome on toP. It was all, all very defined, very clear," he told the newspaper editors. Powell and McClave independently told McDonald that the disk-like device had a "glistening white rounded dome on top and a red conical apron below, circular platform, and moving with its symmetry axis vertical. It had no wings, tail, propellers or jets. No markings or apertures were discerned." Powell estimated that it closed at airspeed of about 200 mph and passed to their right and slightly below their altitude. He estimated the diameter at 20 feet; McClave thought 40 feet. "It was just like looking at a Cadillac," Powell told McDonald. It passed by in a steady motion with no wake, no exhaust and no smoke. Because of the construction of the cockpit windows in the Luscombe Silvaire he was flying, Powell could not easily see the object after it passed to the right. He told the newspaper editors "Miss McClave actually saw it disappear. It never got out of her vision until it all of a sudden disappeared after the aircraft (i.e., the UFO) was on the right hand side." According to McDonald, "both had the distinct impression that after the object passed several tens of degrees aft of the beam it suddenly vanished from sight. To all of my queries as to whether this seeming instantaneous disappearance might have been only a matter of extremely high angular acceleration out of their field of view, both could only reply that they did not have that impression. They felt that it had instantaneously vanished while in full view."

Bruce Maccabee wrote on Feb. 10, 2004.

"Since you have abstracted only about 20% of the acceleration paper I suggest you add in the URL for people who might be interested in reading the whole thing (complete with video analysis that demonstrates extreme acceleration)."

http://www.nidsci.org/articles/maccabee/acceleration.html

Unconventional Flying Objects: How they fly

Paul Hill wrote his generally excellent technical book about thirty years ago before the discovery of anti-gravitating dark energy exotic vacuum in 1999 or so as the dominating stuff of the Universe on the large-scale. In those ancient primitive times physicists used the imprecise term "negative matter" as in the late Robert Forward's papers on "Negative Matter Propulsion" (1988) that is a primitive form of the free-float timelike geodesic zero g-force Alcubierre faster-than-light warp drive already known to top physicists Herman Bondi (British Intelligence) and Yakov Petrovich Terletskii (Soviet Intelligence under Joseph Stalin) in the late 1950's.

"Personally, I don't believe in "Space Drives", but I am willing to keep an open mind and read what people have to say. I have strong doubts that anything will come from gyroscopes or other mechanical devices, but if we can find some new physics (like the negative matter I postulate in the enclosed paper, or the conversion of angular momentum that I postulate in "Spin Drive to the Stars"), then perhaps we can make an end run around the laws of conservation of momentum and energy."
Robert Forward[lxii]

On May 14, 2004, at 9:41 AM, Jack Sarfatti wrote to Gary Bekkum:

Keep me posted on spin off from those two PRAVDA articles.[lxiii] They should stir things up a bit. Awaken The Pundits from their dogmatic slumbers from the hallowed Halls of Ivy to the dark corridors of power at Langley and The Kremlin.[lxiv]The Russians and Ukranians may not have a lot of money, but they still have superior intellects and this sort of stuff does not need expensive machines with huge energy outputs. Eric Davis is wrong about that. It's all micro-nanotech with a fine mesh-phased array of superconducting LC oscillators with really high T_c coils on micron to nano scales s imbedded all over the thin smart computing material of the fuselage. Each little solenoid current controls the local Josephson phase difference between the high T_c SC (superconducting) nano-coil and the vacuum it occupies. Hence, the coherent phased array $\Theta(x,s)$ controls induced $\delta\Lambda_{zpf}(x,s)$ i.e. at scale s in a 2D anyon type UFO fuselage sheet

$$\delta\Lambda_{zpf}(x,s) \sim \chi \sqrt{\rho_{e^+e^-}_{\substack{vacuumODLRO}}(x,s)} \sqrt{\rho_{e^-e^-}_{\substack{2D-anyon}}(x,s)} \cos\Theta(x,s)$$

$$[\rho] \sim \frac{1}{Length^2}$$

$$\Theta \sim \int_{close-loop\to 0} \left(\kappa A_\mu(Anyon) - B_\mu(VacuumODLRO)\right) dx^\mu$$

$$\sim \iint_{loop-area\to 0} d_\nu\left(\kappa A_\mu(Anyon) - B_\mu(VacuumODLRO)\right) dx^\mu \wedge dx^\nu$$

126

Where the gauge invariant relative Josephson phase $\Theta(x,s)$ between the superconductor "control knob" and the vacuum condensate is sensitive to magnetic fluxes, rotations and other Berry phase effects in the usual well known ways and χ is the impedance matching coupling coefficient. The induced exotic vacuum zero point dark energy-stress current density tensor is

$$\delta t_{\mu\nu}(zpf) = \left(\frac{c^4}{8\pi G^*}\right)\delta\Lambda_{zpf}\, g_{\mu\nu}$$

where we want s-regions such that G*(s) >> G(Newton), i.e. increasing Andrei Sakharov's "metric elasticity."[lxv]

$$\Lambda_{zpf} \approx Tr(K)$$

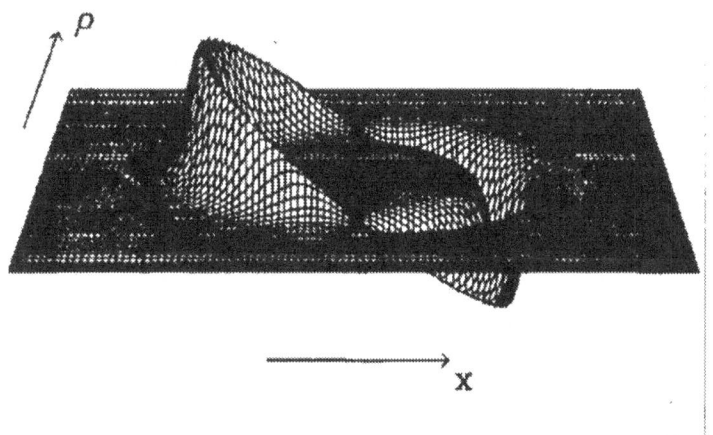

Tr(K) ~ Divergence of Gravimagnetic Field ~ Λ_{zpf}
Components of Gravimagnetic Field 3-vector = g_{oi}

Exempt from Disclosure

By Robert M. Collins & allegedly Richard C. Doty[lxvi] with Tim Cooper (2005)

On Feb 11, 2006, at 1:12 PM, Jack Sarfatti wrote:

"Angleton learned ... the basic tools of counterintelligence ... penetration of the enemu's intelligence services was crucial in order to nudge the enemy into an unreal world – the famous 'wilderness of mirrors' ... doomed to spend their working lives trapped inside the shimmering bars of glancing reflections. It is the place where truth and reality are blinded by deception." Tom Mangold "Cold Warrior" cited by Collins on p. 5.

What makes this book interesting is that it contains a lot of provocative details and documents, which may or may not be true and that Collins was in USAF Intelligence as an officer. Richard Doty was also in USAF Intelligence. Collins claims that Dorothy Kilgallen, JFK and Marilyn Monroe were all murdered because of the alleged UFO coverup. Collins also claims that an ancient Hebrew Bible was found in a crashed saucer. Einstein, John von Neumann and Carl Sagan are alleged to have been involved in trying to reverse engineer the alien technology. I know of no evidence for that allegation. All of these claims, at first sight, makes the book look flaky. James Jesus Angleton, Richard Helms, Allen Dulles, William Colby and other important figures are all cited as major players. Only professional historians will be able to evaluate many of the allegations in the book that I will take no position on since I am not competent to do so. I will only briefly comment on allegations that involve physics.

From Yellow Book to Yellow Cake?
"The extraterrestrials had ... the Yellow Book ... a history ... written by them ... about a crystal that when an ET held it and turned it in certain ways, it showed pictures of their home planet ... Agent Rick Doty indicated that he had seen it but couldn't make it work." p. 11

"The debris from the primary field of the 1947 crash 20 miles southeast of Sorroco, New Mexico was called ULAT-1 ... its unheard of tensile and shear strength. The nuclear ... engine used heavy water and deuterium with an oddly arranged series of coils, magnets and electrodes... related to the use of the isotope H5" p. 15

This contradicts Col. Philip Corso's testimony that there was no visible large-scale engine inside the ship. H5 is nonsense as a propellant. In the free state it only exists for 10^{-21} sec and it is certainly not relevant to weightless geodesic warp drive. I discount the above allegation as disinformation. Finally the claim is made of a power crystal that was tested on board the Space Shuttle that delivers 100,000 watts at 100 amps and never runs out - suggesting zero point vacuum energy. Linda Moulton Howe on the February 10, 2006 Coast to Coast George Noory Show tells a story in which one of the captured living ETs was able to lift 200 lb metal psychokinetically with his mind at a distance. This requires the physics of "signal nonlocality" that violates micro-quantum theory.

The Nazi Bell?

The Nazi's Bell Theorem
"The experiments required large doses of electricity, fed via thick cabling"
(as in the alleged 1943 "Philadelphia Navy Yard Experiment) *"into a chamber hundreds of meters below ground. In this chamber, a bell-shaped device comprising two counter-rotating cylinders filled with mercury or something like it, had emitted a strange pale-blue light. A number of scientists who had been exposed to the device suffered terrible side effects ..."* p. 183, "The Hunt for Zero Point, Nick Cook.

JS: Nick Cook says the Podkletnov had access to Schauberger's original papers captured by the Soviets at end of WWII and that this influenced Podkletnov who also invokes Akimov's and Shipov's "torsion fields" that have been strongly debunked in Russia as shown elsewhere in this book. However, that being said and borne in mind, with my new equations I understand simply how the rotating superconductors of Ning Li and Evgeny Podkletnov can generate anti-gravity in the precise sense of mainstream physics today, i.e. negative quantum pressure for net w = -1 zero point energy density in presence of vacuum coherence. Vacuum coherence is missing totally from Hal Puthoff's ideas for metric engineering, which will not work. I guarantee it. He does not have the right idea. He is missing the key conceptual insight. However, is there another way to couple a control material to the macro-quantum vacuum coherence apart from using the macro-quantum coherence of a rotating superconductor as the control knob? Can we use another kind of macro-quantum coherent "More is different" order parameter apart from that of a high Tc superconductor? Did Schauberger et-al in the "Bell" stumble on such a material that does not need cryogenics? I don't know yet. It is remotely possible, but unlikely. Michael Ibison reports that Puthoff's Austin lab IAS has experimented with a small version of this, but that it does not work.

UFO Metric Engineering

Dr. Harold Puthoff coined the term "metric engineering." Hal, as he prefers to be called, was a US Naval Officer who then worked for the National Security Agency before going to the Stanford Research Institute where he conducted the famous "Remote Viewing" experiments with Russell Targ testing Uri Geller, Ingo Swann, Pat Price and other psychics in a project paid for by the CIA and Department of Defense Intelligence Agencies.

I first met Hal and Russell at SRI in 1973 and that story is told in my book "Destiny Matrix." Hal has held very high USG security clearances and it is well known that he is obsessed with the UFO mystery. Hal, like his co-worker Bernie Haisch, who is an editor of the Astrophysical Journal, both strongly believe in the physical reality of flying saucers. They, with Jacques Vallee, have been active in UFO groups, including the NASA Breakthrough Propulsion Program, some financed by Laurance Rockefeller and the "Howard Hughes" clone Las Vegas Hotel Tycoon Robert Bigelow who owns Bigelow Aerospace Corporation to put Miss Piggy in a Hotel orbiting Earth. (Grin) Since they take the reality of the flying saucers seriously so do I. The two lines of theoretical physics research that Hal has pursued for several decades now, zero point energy and a polarized vacuum model of gravity, is primarily motivated by the quest to understand how the saucers fly. The alleged reality of such advanced alien technology is clearly of immense importance to US National Security and beyond. The recent development in physics shown in NOVA's "Elegant Universe" with Brian Greene, in Stephen Hawking's "The Universe in a Nutshell", Michio Kaku's "Hyperspace" and Igor Novikov's "The River of Time" makes time traveling alien interference in our history more probable not less probable. This is all an aspect of "metric engineering" defined as the practical control of space and time warps. I have discussed this in Paramount Pictures DVD Special Collector's Edition of "Star Trek IV: The Voyage Home" in "Time Travel: The Art of the Possible" and in Learning Channel's "Ultra-Science: Time Travel." The ultimate form of metric engineering is seen in "Q" in "Star Trek" where a Super Mind is able to warp space-time.

We also need a physics of consciousness to see if that fiction can be realized in fact. British Astronomer Royal, Sir Martin Rees who runs the laboratory where Stephen Hawking works and who is the new Master of Trinity College, Cambridge has added the dimension of "Doomsday WMD" to the quest for a metric engineering breakthrough in his Chapter 9 of his pessimistic "Our Final Hour" on the "Ice Nine" "rip-in-space" propagating at the speed of light that could literally destroy our universe. We are now at that turning point in our history where, like Mickey Mouse in Walt Disney's "Fantasia" as "The Sorceror's Apprentice." we could, in our incompetence, destroy the entire Universe just as we are now surely destroying Earth's Biosphere. Therefore, any aliens out there with star gate time travel and warp drive super technology are already here in order to stop us from killing them as well as ourselves. Lev Okun reports that Andrei Sakharov was also very concerned with this issue of exotic "vacuum bubble" WMD.

In the Heart of Dark Energy: Apocalypse Here-Now?

"The Laws of Physics may not only be variable but are almost always deadly. ... Like deadly storms bubbles of extremely hostile environments may propagate through the universe causing destruction in their wake." Lenny Susskind, "Cosmic Landscape" p. 90

Perfect Cosmic Storm – The Vacuum Bubble

Will Israel take out Iran's nuclear weapons facilities in April 2006? I am writing this on December 12, 2005. Global warming seems to be here. We had too many super hurricanes in 2005. New Orleans sank like Atlantis. Will Europe flood and Britain turn to ice? As if that is not enough, there is even more for Woody Allen to worry about:

"Why not go to a lower vacuum here and now? ... It was a very frightening experience which I had when I first thought about these bubbles ... I really shivered ... I told Andrei Sakharov about these bubbles. I vividly remember his reaction. He said: 'Such theoretical work should be forbidden. It's too dangerous."
Lev. B. Okun hep-ph/0112032

See also Okun's "Space-Time and Vacuum as seen from Moscow"[lxvii]

"Experiments that crash atoms together with immense force could start a chain reaction that erodes everything on Earth; the experiments could even tear the fabric of space itself, and ultimate 'Doomsday' catastrophe whose fallout spreads at the speed of light to engulf the entire universe."

P.2, "Our Final Hour", Sir Martin Rees (2003). See Chapter 9 for more details about Kurt Vonnegut's "Ice Nine" type quark strangelets as well as the $G^* \gg G$ theory weakening the string tension space-time stiffness factor I am developing in this book applicable to UFO propulsion. In a recent interview on PBS the 83 year old Vonnegut is increasingly pessimistic that we humans will survive much longer.

"Some new theories, however invoke extra spatial dimensions beyond our usual three; a consequence would be to strengthen gravity's grip, rendering it less difficult to than we previously thought for a small object to implode into a black hole." P. 120.

I already wrote a memo to the US Intelligence about these metrically engineered tiny black holes in 1973 while working at Abdus Salam's ICTP in Trieste, Italy on strong short range "f-gravity" where I explained the universal Regge "string" slope alpha' $(1 \text{ Gev})^{-2}$ of hadronic resonances as rotating micro-geon Kerr-Newman black holes with $G^* \sim 10^{40}$ G(Newton).

Note however:[lxviii]
Review of Speculative "Disaster Scenarios" at RHIC
Authors: R.L. Jaffe, W. Busza, J.Sandweiss, F. Wilczek
Journal-ref: Rev.Mod.Phys. 72 (2000) 1125-1140

"We discuss speculative disaster scenarios inspired by hypothetical new fundamental processes that might occur in high-energy relativistic heavy ion collisions. We estimate the parameters relevant to black hole production; we find that they are absurdly small. We show that other accelerator and (especially) cosmic ray environments have already provided far more auspicious opportunities for transition to a new vacuum state, so that existing observations provide stringent bounds. We discuss in most detail the possibility of producing a dangerous strangelet. We argue that four separate requirements are necessary for this to occur: existence of large stable strangelets, metastability of intermediate size strangelets, negative charge for strangelets along the stability line, and production of intermediate size strangelets in the heavy ion environment. We discuss both theoretical and experimental reasons why each of these appears unlikely; in particular, we know of no plausible suggestion for why the third or especially the fourth might be true. Given minimal physical assumptions the continued existence of the Moon, in the form we know it, despite billions of years of cosmic ray exposure, provides powerful empirical evidence against the possibility of dangerous strangelet production."

Note that strangelets would be real on-mass-shell not the virtual w = -1 off-mass-shell exotic vacuum zero point dark stress-energy/matter density stuff I am talking about in this book. Saucers fly – don't they?

What does mainstream physics tell us about the possibility of metric engineering? The relevant parts of mainstream physics for metric engineering are Einstein's general theory of relativity and micro-quantum mechanics including the "More is Different" emergence of "macro-quantum" coherent superfluid order in ground state and vacuum instabilities developed by condensed matter Princeton physicist P.W. Anderson also seen in high-energy physics in the "Higgs mechanism" and in the "inflationary chaotic cosmology of A. Linde consistent with the "parallel universes" nicely described by Max Tegmark in the May 2003 issue of Scientific American.

Let's start with Einstein's gravity field equation of ~ 1916. Einstein viewed pure space-time geometry like the marble in a statue by Michelangelo with gross matter and radiation as "wood." Einstein's equation, in the 21st Century language, is

Space-Time Curvature = - (Stress-Energy Density)/(String Tension)

Curvature has the physical dimensions of 1/Area, String Tension has Energy/Length and Stress-Energy Density has Energy/Volume. Simple algebra confirms that this way of writing Einstein's equation is correct dimensionally. The "String Tension" is the basic parameter that Ed Witten of the Princeton Institute of Advanced Study uses in his discussions of "M-Theory" that unifies the five limiting cases of "super string theory" in which the elementary leptons, quarks and gauge force bosons are "vibrating strings of pure energy" pulsating in the extra space dimensions of "Calabi-Yau" hyperspace. The ordinary matter and radiation on the right hand side of Einstein's equation are open strings whose ends are stuck to 3-Dim brane worlds of which our universe is one. You can picture a string as a tiny wormhole whose two ends need not be on the same brane world because only then can you not have equal numbers of particles and antiparticles in the same brane world. See Stephen Hawking's parody of Shakespeare's "Tempest" in his "O' Brane New Worlds" in "The Universe in a Nutshell" for a popular discussion of "branes" where there are extra dimensions of space that do not curl up into tiny circles as in the Kaluza-Klein model. They say M theory fell into the end of the 20th Century from the future. Theoretical physicists and mathematicians in our top universities seem to be all going off the deep end. They all believe in a theory whose basic equations they cannot write down. Then they have Balkanized into two street gangs The Loops and The Strings. The Loops admit they cannot really derive Einstein's 1915 classical field equations as a limiting case of their "spin foams" that is an attempt to discretize the space-time continuum. Such a feat would change the dispersion of really high-energy cosmic rays, but so far no luck. No evidence for loops. Not yet at least. Don't get the impression that The Strings have done much better than The Loops – they haven't. One of the chief gadflies here is Stanford's "soft condensed matter" Nobel Prize physicist Robert Laughlin who works with George Chapline. Laughlin makes his case that everything is emergent even the so called fundamental equations as well as quantum measurements in his new book "A Different Universe" that is partly in opposition to Lenny Susskind's book "Cosmic Landscape." Both men are in the same physics department at Stanford and I find both of their books extremely interesting. Laughlin, in the tradition of P.W. Anderson's "More is different" and G. Volovik's "The Universe in a Helium Drop" sees special relativity and the fermions and bosons of the standard model as emergent symmetries and elementary excitations, respectively at a critical point phase transition of hidden variables with infinite coherence length as in critical opalescence that may well be Galilean. Laughlin does not have an emergent gravity theory like I present in this book, but George Chapline has his own theory not the same as mine. Both Laughlin and Chapline have a new "dark star" theory of the black hole event horizon that opposes the information loss theory of Susskind that Stephen Hawking rolled over on at GR 17 in Dublin July 2004. I was there. It's too soon to know who will win.

Dark Energy Stars or Black Holes?

"George Chapline, a physicist at Lawrence Livermore National Laboratory in California, and Nobel laureate Robert Laughlin of Stanford University ....Last week at the 22nd Pacific Coast Gravity Meeting in Santa Barbara, California, Chapline suggested that the objects that till now have been thought of as black holes could in fact be dead stars that form as a result of an obscure quantum phenomenon. These stars could explain both dark energy and dark matter. ... Chapline and Laughlin found some answers in an unrelated phenomenon: the bizarre behaviour of superconducting crystals as they go through something called "quantum critical phase transition" (New Scientist, 28 January, p 40). During this transition, the spin of the electrons in the crystals is predicted to fluctuate wildly, but this prediction is not borne out by observation. Instead, the fluctuations appear to slow down, and even become still, as if time itself has slowed down. ...The team's calculations show that the vacuum energy inside the shell has a powerful anti-gravity effect, just like the dark energy that appears to be causing the expansion of the universe to accelerate... the anti-gravity should spit some of the remnants back out again. Also, quark particles crossing the shell should decay by releasing positrons and gamma rays, which would pop out of the surface. This could explain the excess positrons that are seen at the centre of our galaxy, around the region that was hitherto thought to harbour a massive black hole. Conventional models cannot adequately explain these positrons, Chapline says. ...The most intriguing fallout from this idea has to do with the strength of the vacuum energy inside the dark energy star. This energy is related to the star's size, and for a star as big as our universe the calculated vacuum energy inside its shell matches the value of dark energy seen in the universe today. These small dark energy stars would behave just like dark matter particles: their gravity would tug on the matter around them, but they would otherwise be invisible."
Written by Zeeya Merali The complete article appears in New Scientist: 11 March 2006

Note that in my theory all on-mass-shell quarks and leptons are tiny micro-geon Bohm hidden variables of spatially extended shells of electric, weak and strong charges glued attractively glued together by positive pressure zero point vacuum energy. They appear to shrink under high magnification from the extreme warping of space provided by the momentum transfer of the probe itself. What is not clear in the above popular article is how the repulsive negative pressure dark energy inside the quantum critical shell flips over to an attractive gravity field outside shell or what is usually assumed to be the surface of a "black hole."

One weakness in this mainstream civil war is that all parties, except me, assume that micro-quantum theory's linearity and unitarity are sacrosanct at the black hole surface. This may not be so. If "more" really is different, then the Landau-Ginzburg equation replaces the Schrodinger equation with a new set of interpretive rules that are local rather than nonlocal, that are nonunitary rather than unitary, that are non-probabilistic rather than probabilistic, and that, counter-intuitively, replace nonlocal micro-quantum signal locality with local macro-quantum signal nonlocality. There is no "unitary S-matrix" here to have to satisfy. The presponse data of living matter, the odd paranormal Skinwalker evidence and the chilling UFO flying saucer evidence are all germane although the mainstream physicists continue to reject those data points out of peer pressure.

Another equivalent way to look at Einstein's 1916 equation from our post-modern 2004 point of view is as a perfect balance on The Scale of Cosmic Justice peeking through the blindfold. That is, the stress-energy density "tensor" of pure marble geometry is

Stress-Energy Density Tensor of Pure Geometry
= (String Tension)(Space-Time Curvature)

Einstein's equation is then like the static equilibrium balance of forces in architecture in which the sum of all the contributions to the stress-energy density tensor add up to a perfect zero! In this simplest of cases

Stress-Energy Density Tensor of Geometry + Stress-Energy Density of Matter = 0

All physically real objects are either tensors, spinors or twistors in the theory of relativity. This is because the coordinate map is not the physical territory. The form of the laws of physics must be the same locally no matter how the coordinate map is morphed. To this we must add the Einstein Equivalence Principle or EEP, which says that even in a curved space-time, we can use the special theory of relativity locally for a special class of LIF observers to describe what is happening to a good approximation. The EEP is only meant to apply in this approximate way and it will break down if one is falling into a black hole singularity of if the microscope magnification is so large that quantum gravity zero point energy density fluctuations in the space-time geometry itself get large. There may also be torsion fields not included in Einstein's 1916 "geometrodynamics" that may require a modification of the EEP. The spinor is a square root of a tensor and the Penrose twistor is a spinor in a complex space-time. The physics of point particles in real space-time when extended to complex and hyper-complex matrix space-time is in reality the physics of extended strings and membranes in real space-time. Hawking's "imaginary time" and its connection to inverse temperature is part of that same story.

Like the "Underground Stream" of the old Hermetic Alchemists and Fulcanelli, the key concept for metric engineering in the new "Techgnosis" is the Einstein esoteric current! Einstein's field equation is always the statement that the sum of all the stress-energy density tensors when put on the same side of the equation all balance out to exactly zero. We then take a "covariant divergence" of the equation to get the Einstein currents, which are conserved when added together. The covariant divergence depends on something called a "connection field", which tells us how to "parallel transport" tensors and spinors along paths or "histories" in space-time and beyond including the extra dimensions of hyperspace. Every time a gauge force field is added, like a torsion field, there is an additional piece added to this connection field. Einstein's 1916 theory is a degenerate case of the bigger unified field theory just like a circle is a degenerate case of an ellipse. An ellipse has two centers or "foci." When the two foci merge together the ellipse degenerates into a circle. Einstein's 1916 theory in the form of the "Bianchi identities" forbids practical metric engineering because it has an impenetrable barrier completely separating the marble geometric current from the wood matter current.

Both kinds of currents must be able to intermingle to transform into each other in order to have "soft" practical metric engineering of Star Gate Time Travel traversable wormholes and weightless superluminal warp drives using the anti-gravitating exotic vacuum "dark energy" with negative zero point exotic vacuum pressure that is equal in magnitude, but opposite in sign, to the zero point energy density. The new Type Ia supernovae data showing that our brane world Universe is accelerating in its rate of expansion, i.e. speeding up rather than slowing down. Indeed approximately 3/4 of all the stuff of our Universe on the large scale is made out of "dark energy."

Therefore, the idea of ancient traversable Star Gate wormholes stabilized by dark energy is not so far fetched. The brute force approach to metric engineering taken by Hal Puthoff now for at least two decades can never work because space-time is too stiff to bend directly with electromagnetic field energy density to do anything worth doing because the string tension is too large. In contrast, we can used the quantum interference of the Einstein stress-energy density currents via the Bohm-Aharonov-Josephson effect for practical metric engineering provided that the pure geometric currents are not separately conserved like they are in Einstein's original theory. One way to do that is to "bottle" or "harness" the "dark energy" as General Douglas Mac Arthur precognitively remote viewed in his "Duty, Honor, Country" Farewell Speech to the Cadets at West Point in 1962 in a speech made even more remarkable for its reference to the coming war in space with extra-terrestrials. Michael Turner, a professor of physics at the University of Chicago, says in April 2003 Physics Today that it is impossible to "bottle dark energy" and if Einstein's 1916 theory is the final theory he is correct. However, the real UFO evidence that drives Hal Puthoff like Captain Ahab after The White Wale, Moby Dick, is the evidence that suggests Professor Turner will be proved wrong in that prediction.

Mesoscopic Exotic Vacuum Objects?

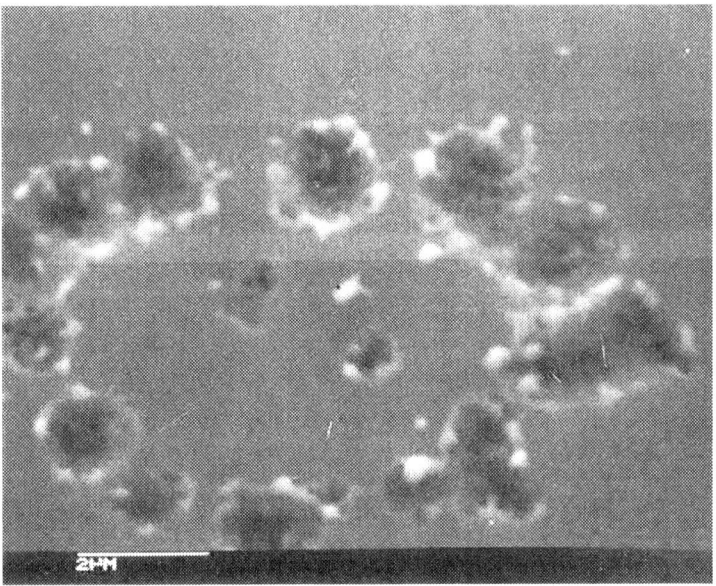

From Ken Shoulders' laboratory: these are allegedly large numbers of electrons glued together by a core of strong positive zero point energy pressure.

There is the thermodynamic-mechanical (i.e. electric) face and the gravitational face.

In thermodynamics

$$P = - dU/dV$$

U = internal energy

Consider a piston in a gas-filled cylinder, you need to do work to compress the gas, U increases as V decreases when P is positive. If P is negative, you have to do work to expand the gas. The latter suggests stretching spring and a confinement effect, so if that is so, how does negative vacuum pressure cause the universe to accelerate in its expansion? That's where the GR gravity effect comes in. In Einstein's GR the negative pressure provides its own repulsive gravity force that does the work to expand space. In the large scale $\Lambda \sim$ zero point energy density is constant and uniform so that the total global internal energy of the vacuum of the universe is increasing as space expands from the work done by the repulsive gravity "force" (in Newtonian terms).

However, in both the Galactic Halo on large scale and inside a thin spherical shell of charge on the small scale we want the zero point pressure to be positive with $w < - 1/3$ so that the strong short-range attractive gravity induced directly from those virtual vibrations inside the vacuum hold the self-charge together.

The big difference between Hal Puthoff's explanation of the EVOs and mine is that

Hal assumes a positive zero point energy strictly mechanical pressure outside the charged shell with zero pressure inside the shell. In contrast, I assume a positive pressure inside the shell with almost zero pressure outside the shell. Actually there is a small amount of negative pressure outside the shell of charge that is 73% of the stuff of the universe, i.e. 73% of the critical density ~ 10^{-29} grams/cc. Hal's model is no good because it gives a huge cosmological constant of

$$\Lambda_{Puthoff} = \left(\frac{mc}{h}\right)^2 \sim 10^{22} cm^{-2} \qquad (0.29)$$

In contrast, my model is consistent with observation

$$\Lambda \sim \left(\frac{H(t)}{c}\right)^2 \sim 10^{-56} cm^{-2} \sim 10^{-78} \Lambda_{Puthoff} \qquad (0.30)$$

Furthermore, Hal has his quantum pressure to be one third of the zero point energy density, i.e. w = +1/3. This contradicts Einstein's general relativity's equivalence principle, which demands that the exotic vacuum quantum pressure is the negative of the zero point energy density, i.e. w = -1.

On Dec 12, 2005, at 9:42 PM, Jack Sarfatti wrote:

Hal used Bryce Dewitt's tensor with an effective w = +1/3, well that is only for anisotropic parallel plates case as Matt Visser explained in his book "Lorentzian Wormholes." There the longitudinal pressure is 3x the 2 transverse pressures and is same sign as energy density. It obviously does not apply to the spherical shell geometry where obviously the 3 pressures are the same in the source stress-energy tensor. That is, for parallel plates +1 - 1 - 1 + 3 on the trace on the diagonal of the source stress tensor obeys

1 + 3w = +2

w = +1/3

In contrast in a spherical shell geometry, the diagonal entries in the source tensor are proportional to

+1 - 1 -1 -1

1 + 3w = -2

w = -1

We need negative ZPF energy density of positive pressure to hold the charges together. Inside we want positive pressure with zero pressure outside (actually slightly negative). That is like the Galactic Halo its a dark matter core not a dark energy core.

Finally, my model solves the old problem of the stability of the classical electron raised by Lorentz et-al more than 100 years ago and never solved.

Hal and I both use the same basic formula for the surface area of the EVO

$$N\left(\frac{h}{mc}\right)^2 \sim 4\pi r_{EVO}^2 \qquad (0.31)$$

For an EVO with charge Ne and *apparent* radius r_{EVO} in the sense of the Schwarzschild radial coordinate. In my model the actual interior radius is much larger than r_{EVO} because of the large short-scale space warp caused by the large positive zero point quantum pressure in the interior of the EVO, which is a mesoscopic charged Wheeler geon. The positive quantum pressure in the interior of the charged shell sucks in the repulsive Coulomb energy preventing the explosion of the EVO at least temporarily. The large EVOs have many low-lying collective modes that destabilize it compared to a single spinning electron N = 1 needing a modified Kerr-Newman metric with a large effective $G^* \sim 10^{40} G_{Newton}$.

Theoretical Model of the EVOs

Einstein's classical general theory of relativity combined with Heisenberg's quantum uncertainty principle shows that any zero point energy density in the vacuum has a direct gravity influence that can either be attractive or repulsive depending on the local value of the exotic vacuum coherence field Ψ. We are interested in the potential energy per unit test mass. Consider a rotating thin shell of electric charge

$$V = \frac{J^2}{Nm^2r^2} + \frac{(Ne)^2}{Nmr} - c^2\Lambda_{zpf}r^2 \tag{1.1}$$

$$-\frac{\partial V}{\partial r} = -\frac{J^2}{Nm^2}\frac{\partial}{\partial r}\frac{1}{r^2} - \frac{(Ne)^2}{Nm}\frac{\partial}{\partial r}\frac{1}{r} + c^2\Lambda_{zpf}\frac{\partial}{\partial r}r^2$$
$$= +2\frac{J^2}{Nm^2}\frac{1}{r^3} + \frac{(Ne)^2}{Nm}\frac{1}{r^2} + 2c^2\Lambda_{zpf}r \tag{1.2}$$

The equilibrium critical point where this derivative vanishes is

$$-\frac{\partial V}{\partial r} = +2\frac{J^2}{Nm^2}\frac{1}{r^3} + \frac{(Ne)^2}{Nm}\frac{1}{r^2} + 2c^2\Lambda_{zpf}r = 0$$
$$+2\frac{J^2}{Nm^2} + \frac{Ne^2}{m}r + 2c^2\Lambda_{zpf}r^4 = 0 \tag{1.3}$$
$$\Lambda_{zpf} < 0$$

Stability requires

$$\frac{\partial^2 V}{\partial r^2} > 0 \tag{1.4}$$

In the case of a single electron as hollow thin shell of charge with a lot of negative zero point energy density inside it, we need to add the Casimir term that I ignored above in the large charge cluster $N \sim 10^{11}$ to 10^{16} limit. The single electron Bohm hidden variable micro-geon effective potential in the weak field limit is

$$V_e = \frac{\varsigma\hbar c + e^2}{mr} + \left(\frac{\hbar}{mr}\right)^2 + c^2\Lambda_{zpf}r^2 \tag{1.5}$$

Shall we try to change rest mass? NO!

Now to un-garble a key quote by Nick that appears at least twice.

"I ... found five possible pathways to anti-gravity: manipulating an object's mass and/or inertia; exploitation of the zero point energy field; perturbations of the space-time continuum; faster-than-light travel; and gravity shielding." p. 117 This idea has been propounded at STAIF meetings published by the American Institute of Physics. This shows that many military engineers do not know this kind of physics very deeply. The idea is fundamentally confused as far as anti-gravity propulsion is concerned. The idea is irrelevant because it shows a fundamental misunderstanding of Einstein's, indeed, even Galileo's equivalence principle that the inertia completely cancels out of the problem in timelike geodesic motion with zero g-force and small tidal curvature force gradients. Indeed, even Galileo knew this in rudimentary form. In terms of high school Newtonian physics:

Weight = W = Inertia x g-field of gravity = mg, using **bold face** for vectors in space:

$$\mathbf{W} = m\mathbf{g}$$

Any real anti-gravity propulsion leaves m alone, it merely changes g! In fact it makes g = 0 even though the self-controlled timelike geodesic path of the saucer can make U turns that normally might look like 1000 g(surface of Earth) to the outside observer. W = 0 for all objects inside the saucer when Nick Cook's "anti-gravity switch" is "ON." m does not change at all! Even if you could change the rest mass m you do not want to do it because you then change e/m, and h/mc and you destroy the delicate balance for the stability of matter as explained by Sir Martin Rees in "Just Six Numbers", by Paul Davies in "Accidental Universe" and by John Barrow and Frank Tipler in "The Anthropic Cosmological Principle." James Woodward, a physics professor at Cal State in the LA area has an actual machine that he claims can change the rest mass using Mach's Principle. I have no idea if his machine works, but if it does then it is potentially quite dangerous for the reasons stated. In principle one could have propellantless propulsion this way but it would be non-geodesic with g-forces and time dilation. This is not the geodesic warp drive without any g-forces for the occupants nor any time dilation as in Alcubierre's toy model.

Casimir Warp Drive? Not likely.

The ratio of Casimir electrical attraction between the plates to the opposing simultaneous Casimir gravity repulsion is the area of the plates multiplied by the Compton wavelength of the plates divided by the Planck area multiplied by the separation of the plates. This is large compared to 1, i.e. the Casimir induced gravity repulsion is relatively weak in all practical devices. Note that the area of the plates divided by the Planck area is essentially the number of Bekenstein-Hawking BITS of the plates.

Nuclear Isomer Gamma Ray Weapons?

History of the gamma ray laser concept 1963 at Cornell with Hans Bethe[lxix]

"Isomer weapons would not involve transmutation of nuclear species ... Nuclear isomers are excited states that decay ... mostly by gamma radiation ... only a few have half lives longer than a day."
"Conflicting results on a Long-Lived Nuclear Isomer of Hafnium..." Physics Today, May, 2004 pp. 21-24.

I worked on the concept of a nuclear gamma ray laser using isomers with Hans Bethe back at Cornell in 1963. It was my idea that I got when I was at Tech/Ops/Mitre in Burlington, Mass (USAF/CIA contractor) in George Parrent Jr.'s group with British physicist Brian Thompson. After a short while Bethe discouraged the idea, but I think maybe he took it "black"? I do not know. That was when I was back at Cornell with Lenny Susskind and Johnny Glogower working on quantum phase/time operator problem that I also got from working with George Parrent Jr who was a student of Emil Wolf's. Their main project was imaging targets on ground from high-level U2s and satellites. Paul Roman from BU was part of that on the periphery and George wanted me to stay at Tech/Ops and work with Paul Roman at BU on Ph.D. I had also turned down an offer from Leonard Schiff in a phone call to go to Stanford and a job offer from Naval Surface Weapons Lab in California.

Report to The American Physical Society of the study group on science and technology of directed energy weapons
My nuclear isomer idea was one of the reasons Ron Bullough, head of Theory Division AERE Harwell, invited me there in 1996. George Chapline, working for Edward Teller, actually exploded an underground nuclear weapon to test for stimulated emission of a coherent beam of high-energy photons. They were X-rays however not gamma rays.[lxx]

On Mar 12, 2006, at 5:35 PM, Sharon Weinberger wrote:
"For those of you I've interviewed or spoke with informally over the past couple years about the "isomer bomb" and related controversies, or have expressed interest in the subject, I wanted to provide the following brief information.

The book, "Imaginary Weapons: A Journey Through the Pentagon's Scientific Underworld" is scheduled for publication on June 13, 2006.

The book deals primarily with the controversy over the isomer bomb, although it draws on a number of interviews I've done over the past couple years on other controversial and/or fringe science issues.

I should get a limited number of copies that my publisher, Avalon, is providing me, and I'll try to distribute those generously to those of you who read drafts of the book and/or contributed substantially to the content, for which I'm eternally grateful."

My behind-the-scenes role in getting President Reagan to opt for SDI, perhaps another "Imaginary Weapon", perhaps not, is described in my autobiography "Destiny Matrix" and is alluded to in "The Buttoned Down Bohemians" San Francisco Chronicle Sunday Magazine (1986) about the Institute for Contemporary Studies the leading policy think-tank for Reagan set up in San Francisco by Cap Weinberger and Ed Meese while Reagan was still Governer of California

"Since he first came to the Bay Area in the mid-1970s, physicist Jack Sarfatti has been a provocative presence in local intellectual life. Leaping from North Beach cafes to *leading policy think-tanks*, he has cut a broad intellectual swath, challenging the preconceptions of poets, political thinkers and physicists alike. With a background in quantum theory, he claims to break new ground in scientific understanding of the eternal questions: 'Who are we? Where do we come from? Where are we going?' In this interview, he discusses the breakdown of a paradigm that, for centuries in the West, has viewed science and humanistic thought as irrevocably separated." Stephen Schwartz in the San Francisco Chronicle Sunday Magazine of August 17, 1997.[lxxi]

Keith Brueckner UCSD & DOD JASONS

The other was my idea for using laser pressure to confine hot fusion plasma. I later was in Keith Brueckner's group at UCSD and he worked on that problem. Brueckner part of the JASONs who used to meet at UCSD. Also George Chapline Jr actually made a nuclear laser for Teller years later (triggered by an A Bomb). George got me my job at SDSU and he used to come down to UCSD from Cal Tech all the time. Greg Benford was part of this little circle in La Jolla. See his novel "Timescape" about future to past signals - another odd synchronicity!

Harwell, UKAERE 1966

It was at Harwell that I wrote the paper with A. M. Stoneham (later head of Theory Division at Harwell) on the "Goldstone Theorem and the Jahn-Teller Effect" in Proc London Physical Society cited in AIP "Resource Letter on Symmetry in Physics"

This paper (1967) may have been first application of Goldstone spontaneous broken ground state symmetry to a practical problem of interest to nuclear reactor engineers. I got interested in broken symmetry from going down to Imperial College, London to meet Abdus Salam & Kibble. Stoneham, based on what I told him of my London adventures, saw how to apply it to the Jahn-Teller effect in crystal distortions.

American Military Investigated Anti-gravity Weapons?

"Government Budget Document evidence that the United States Department of Defense actively pursued reports of anti-gravity effects The United States Government Official Website[lxxii], allows a rapid search of official government documents available over the internet. A quick search for Dr. Ning Li, the physicist that predicted an anti-gravity effect from her theory, and the key word "contract" leads to an official US Government Department of Defense budget document: The document describes contract number DAAH01-01-9-R001, titled "Gravito-Electro Magnetic Superconductivity Experiment", awarded by the US Army Aviation and Missile Command, to Dr. Ning Li's company AC Gravity, LLD"[lxxiii]

"New means of propulsion, ways of controlling missiles and gun launched munitions, lowering of the effective weight of tanks and heavy vehicles, deflecting incoming missiles, including ICBM's, are listed as among the military applications for this technology. This document supports Jane's Defence Weekly contributor Nick Cook's exposure of a similar project at the American aerospace contractor Boeing. Previously Nick Cook showed the BBC and Jane's documents as proof that Boeings' Phantom Works facility were investigating the Russian scientist Evgeny Podkletnov's reports of repulsive antigravity from spinning superconducting disks."

Gary S. Bekkum[lxxiv]
Starstream Research
USA

Exotic WMD?

"Two big stories from the world of physics may portend the arrival of new weapons of mass destruction far more powerful and compact than atomic bombs.

In recent years it has been discovered that our universe is being blown apart by a mysterious anti-gravity effect called "dark energy." Mainstream physicists are scrambling to explain this mysterious acceleration in the expansion of the universe. Some physicists even believe that the expansion will lead to "The Big Rip" when all of the matter in the universe is torn asunder - from clusters of galaxies in deep space down to the tiniest atomic particles. The universe now appears to be made of two unknowns - roughly 23% is "dark matter", an invisible source of gravity, and roughly 73% is "dark energy", an invisible anti-gravity force. Ordinary matter constitutes perhaps 4 percent of the universe. Recently the British science news journal "New Scientist" revealed that the American military is pursuing new types of exotic bombs - including a new class of isomeric gamma ray weapons."

That was an original idea of mine in 1963 at Cornell and I discussed it with Hans Bethe.

That is one of the reasons Ron Bullough invited me to Harwell in 1966. No doubt others thought of it but probably later. I thought of it while at Tech/Ops in Lexington, Mass working for George Parrant Jr.

"Unlike conventional atomic and hydrogen bombs, the new weapons would trigger the release energy by absorbing radiation, and respond by re-emitting a far more powerful radiation. In this new category of gamma-ray weapons, a nuclear isomer absorbs x-rays and re-emits higher frequency gamma rays. The emitted gamma radiation has been reported to release 60 times the energy of the x-rays that trigger the effect."

Gamma-ray weapons could trigger next arms race 19:00 13 August 03
"Exclusive from New Scientist Print Edition. Subscribe and get 4 free issues.

An exotic kind of nuclear explosive being developed by the US Department of Defense could blur the critical distinction between conventional and nuclear weapons. The work has also raised fears that weapons based on this technology could trigger the next arms race. The explosive works by stimulating the release of energy from the nuclei of certain elements but does not involve nuclear fission or fusion. The energy, emitted as gamma radiation, is thousands of times greater than that from conventional chemical explosives.

*The technology has already been included in the Department of Defense's Militarily Critical Technologies List, which says: "Such extraordinary energy density has the potential to revolutionise all aspects of warfare." Scientists have known for many years that the nuclei of some elements, such as hafnium, can exist in a high-energy state, or nuclear isomer, that slowly decays to a low-energy state by emitting gamma rays. For example, hafnium-178m2, the excited, isomeric form of hafnium-178, has a half-life of 31 years. The possibility that this process could be explosive was discovered when Carl Collins and colleagues at the University of Texas at Dallas demonstrated that they could artificially trigger the decay of the hafnium isomer by bombarding it with low-energy X-rays (**New Scientist** print edition, 3 July 1999). The experiment released 60 times as* *much energy as was put in, and in theory a much greater energy release could be achieved."[lxxv]*

I was thinking in 1963 of a gamma ray laser pumping a nuclear isomeric transition. Bethe at the time said it wouldn't work and basically discouraged me working on it.

Bekkum continued:

"In the summer of 2000 I contacted Nick Cook, the former aviation editor and aerospace consultant to Jane's Defence Weekly, the international military affairs journal. Cook had been investigating black budget super-secret research into exotic physics for advanced propulsion technologies. I had been monitoring electronic discussions between various American and Russian scientists theorizing about rectifying the quantum vacuum for advanced space drive. Several groups of scientists, partitioned into various research organizations, were exploring what NASA calls "Breakthrough Propulsion Physics" - exotic technologies for advanced space travel to traverse the vast distances between stars. Partly inspired by the pulp science fiction stories of their youth, and partly by recent reports of multiple radar tracking tapes of unidentified objects performing impossible maneuvers in the sky, these scientists were on a quest to uncover the most likely new physics for star travel. The NASA program was run by Marc Millis, under the Advanced Space Transportation Program Office (ASTP). Joe Firmage, a Silicon Valley entrepreneur, who at age 28 had risen to CEO of a three billion dollar internet firm, began to fund research in parallel with NASA. He hired a NASA Ames nanotechnology scientist, Creon Levit, to run the "International Space Sciences Organization,"

Joe did that because I suggested it. I introduced Creon to Joe.

"Cook was intrigued by the apparent connections between various private investors, defense contractors, NASA, INSCOM (American military intelligence), and the CIA. While researching exotic propulsion technologies Cook had heard rumors of a new kind of weapon, a "sub-quantum atomic bomb", being whispered about in the dark halls of defense research."

I think that must have come from me regarding J. P. Vigier's "tight atomic states" with experiments in Beograd, Serbia by Z. Maric (same name as Einstein's wife, Mileva Maric) and G. Dragic. But how did Cook hear about that? We brought Vigier to ISSO in San Francisco several times along with physicist Gennady Shipov from Moscow. That story with photographs of Vigier and the group is in my autobiography "Destiny Matrix." Dragic A, Maric Z, Vigier JP; Phys. Lett. A 265 (2000) 163. "New quantum mechanical tight bound states and 'cold fusion'." Creon Levit and Vigier met with Maric in Budapest, Hungary in 2000.

Maric and Dragic test Vigier's Cold Fusion theory of "Tight Atomic States" in Beograd: The sub-Atomic Bomb?

Bekkum who is one of my ISEP "on line virtual students" continued:

"Sub-quantum physics is a controversial re-interpretation of quantum theory, based on so-called pilot wave theories, where an information field controls quantum particles. The late Professor David Bohm showed that the predictions of ordinary quantum mechanics could be recast into a pilot wave information theory. Recently Anthony Valentini of the Perimeter Institute has suggested that ordinary quantum theory may be a special case of pilot wave theories, leaving open the possibility of new and exotic post-quantum technologies. Even thought rumors of a sub-quantum bomb may be purely fantasy ..."

It's not fantasy. It might not work. Randall Mills at Blacklight Power is making similar claims. However, physicists are not able to read his theory papers. However Blacklight Power has gotten tens of millions of dollars and are making claims of empirical success. I remain skeptical. Black Light Power may be an example of what Robert Laughlin calls the "Dark Side" of research in his book "A Different Universe." Maric and Dragic in Beograd, while not ostensibly trying to make a weapon by any means, were trying to test Vigier's basic theory of the spatially extended electron, which I think is basically a correct idea. It fits my own ideas including why the electron appears to shrink to less than 10^{-16} cm under high resolution imaging (i.e. scattering) and how the electric charge distribution is contained by the strongly attractive short-range zero point stress-energy density $w = -1$ exotic vacuum core only roughly one Fermi (10^{-13} cm) across. This solves the Abraham-Becker-Lorentz-Poincare compensating stress problem of 100 years ago. My new explanation of this long-standing mystery that stumped even Richard Feynman, only works in the Bohm "hidden" or "extra" variable interpretation. That is, a classical spatially extended electric charge distribution is unstable. It explodes under its own self-repulsion. This is why physicists had to postulate a point electron because they did not understand that the strong short-range gravity attraction of the positive zero point pressure in the "dark matter phase" of exotic vacuum would hold the charge together. As Herbert Frohlich told me at UCSD in La Jolla in 1966 "the basic thing wrong with physics is the idea of the point electron." The bad idea of the point electron gives the infinite energy in quantum electrodynamics. Richard Feynman told me in his office at Cal Tech in 1968 that "infinite renormalization" is a bait-and-switch shell game like shoddy used car salesmen use, and that it was "a scandal in physics that no one could do better" than what he had done. No wonder all our top theoretical physicists did not balk at sharing Werner Erhard's fine Cuban cigars and liquour in his attic at "The Brown House" on Franklin Street in San Francisco. (Grin) Physicists did not know 100 years ago that 1/3 or so of the universe was this kind of exotic vacuum. For example, there is a huge sphere (AKA "galactic halo") of exotic vacuum of $w = -1$ positive pressure that holds our galaxy together preventing our solar system from going off into space on its own. This sphere looks like $w = 0$ "cold dark matter" from our vantage point. What works on this large scale also works on the small scale of the single electron (and all the charged lepto-quarks). A neutrino has some mass and is simply a micro-geon of "Mass without mass" pure zero point energy with positive pressure. One can plausibly metric engineer exotic vacua into artificial black holes and traversable wormhole star gates as well as weightless warp drives as we allegedly see in the seemingly impossible maneuvers of the flying saucers. A real particle-antiparticle pair is a Wheeler wormhole with at least two mouths (like 2D spherical surfaces in 3D space) and quantized electro-weak-strong Faraday lines of gauge force forming closed loops with no free ends threading the mouths. The fact that there is very little naturally occurring anti-matter in our universe is direct evidence for at least one parallel universe next door where all the other twin wormhole mouths are attached. This is consistent with Jacques Vallee's "Fastwalker." Furthermore, none of the dark matter detector experiments now planned or underway will ever click with the right stuff if my ideas here are correct. This is a matter of fundamental principle.

"There is no question that physicists seriously contemplate a phase transition in the quantum vacuum as a real possibility. The quantum vacuum defies common sense, because empty space in quantum field theory is actually filled with virtual particles. These virtual particles appear and disappear far too quickly to be detected directly, but their existence has been confirmed by experiments that demonstrate their influence on ordinary matter."

A major component of the physical quantum vacuum consists of virtual electrons frothing and bubbling, like The Brew of the Three Witches in Shakespeare's "Macbeth," at the foamy Fermi surface edge of the Dirac negative energy sea. This is because of the Pauli exclusion principle that only none or one electron per quantum state. A virtual electron pops out of the vacuum's Fermi surface leaving a hole behind. The hole is the virtual positron. The result is a "virtual electron-positron pair." However, the virtual electron and the virtual positron attract because they have opposite charges and they are exchanging virtual photons. Therefore, some of them form a more stable bound state. An enormous number of these virtual pairs Bose-Einstein condense into the same center of mass quantum wave packet of the single pair forming the "Vacuum Coherence Field" (AKA "Inflation Field"). This is a dynamic steady state non-equilibrium "dissipative structure" of detailed balance in which there is a continual inflow and outflow of virtual pairs into and out of this giant quantum or "macro-quantum" "superfluid." That is, virtual pair bound states are continually created and destroyed from and into the electrically neutral "ionized plasma" of broken virtual pairs which is the zero point dark energy/matter "normal fluid" component of the exotic vacuum that sustains the robust "generalized phase rigidity" of the macro-quantum vacuum coherence field. This P.W. Anderson "phase rigidity" explains Andrei Sakharov's "metric elasticity" or "space-time stiffness" or "string tension." Essentially this is a vacuum phase transition, similar to the BCS phase transition from a normal metal to an electrical superconductor, from the globally flat micro-quantum electrodynamic vacuum without any gravity at all to the curved macro-quantum electrodynamic vacuum with emergent gravity. Einstein's field equation of general relativity can be derived from the phase wiggles and ripples in the robust stable macroscopically occupied center of mass quantum wave packet of the bound state of the virtual electron-positron pair. The exotic vacuum dark energy and dark matter are simply the amplitude wiggles and ripples of this same virtual pair quantum wave packet. The wave packet spreads over the entire 3D space of the post-inflationary bubble on which our Hubble-horizoned universe is located along with an infinity of parallel universes next door as Max Tegmark explains in May 2003 Scientific American. If Lenny Susskind's world hologram idea is correct, take the surface area of the expanding Hubble sphere that is the causal retarded boundary of 3D space of our past light cone at Earth and divide it by the quantum of area. That gives us the number of Bekenstein-Shannon c-bits and explains the irreversible "arrow of time" (AKA Second Law of Thermodynamics) of increasing thermodynamic entropy in terms of the dynamical expansion of the 3D space of the universe. Lenny Susskind calls this kind of idea "DeSitter Space", which is a large-scale homogeneous isotropic limiting case of what I am pointing to on all scales. However, it is pretty obvious from P.W. Anderson's "More is different" that the world hologram idea of t'Hooft and Susskind is wrong in principle because it assumes that the unitary rules of micro-quantum S-Matrix theory apply to large objects like stars, galaxies and black holes. Vacuum ODLRO introduces nonunitarity in an essential way so that Hawking, in my opinion, was right the first time.

As Above So Below
"Such research should be forbidden!"
Sakharov to Lev Okun

Wagging The Tiger's Tale

It's already too late. Pandora's Box is open. Schrodinger's Cat has jumped out of it. This Kitty has grown up to William Blake's "Tyger, Tyger shining bright in the darkness of the night" and I am holding it by the tail. Hold on to your hats for a bumpy ride until we switch on zero g-force weightless warp drive!

"Tiger, tiger, burning bright
In the forests of the night,
What immortal hand or eye
Could frame thy fearful symmetry?

"In the early 1970's Soviet physicists were concerned that the vacuum of our universe was in fact only one possible state of empty space. The fundamental state of empty space is called the "true vacuum." Our universe was considered to reside in a "false vacuum", protected from the true vacuum by "the wall of our world." A change from one vacuum state to another is known as a phase transition. This is analogous to the transition between frozen and liquid water. Lev Okun, a Russian physicist and historian recalls Andrei Sakharov, the father of the Soviet hydrogen bomb, expressing his concern about research into the phase transitions of the vacuum. If the wall between the vacuum states was to be breached, calculations showed that an unstoppable expanding vacuum bubble would continue to grow until it destroyed our entire universe! Sakharov declared "Such research should be forbidden!" since there was always the possibility that an experiment might accidentally trigger a vacuum phase transition."

British Astronomer Royal, Sir Martin Rees, Master of Trinity College, and Director of the Cambridge University Institute of Theoretical Astronomy on Madingley Road where Stephen Hawking works discusses all this in Chapter 9 of his important book "Our Final Hour."

"Could the wall of our universe be breached from within? The amount of energy required to punch a hole through the wall appeared to be enormous, and no known natural physical phenomena, even the most energetic, had punched through either. A recent report commissioned to examine potential dangers at the Large Hadron Collider, one of the next generations of particle accelerators, concluded that we were safe, to the best of our existing knowledge. Others are not so certain, however. At least one of the Russian physicists I had corresponded with was said to have been a former associate of Andrei Sakharov. He strongly hinted at new theories the Russians had developed which allow for the manipulation of the fundamental constants of nature, but he never revealed more than a sketch of his ideas. He claimed that a breakthrough was within reach, perhaps within five years ... Recent theoretical explorations may suggest another approach to the physics of the vacuum. The invisible gravitating dark matter could be the other side of the invisible dark energy coin, and that suggests the possibility of manipulating the vacuum for energy release."

Now this is my original idea that you got from our communications over the past few years. I am the only physicist in the world today, as far as I know who has suggested this and has already published it in my two books of 2002 so it's in the official record at the Library of Congress.

"If a controllable parameter could be found to mediate the balance between the invisible dark forces, the result would unleash the vacuum energy of creation in all of its awful power and majesty. If it were possible to control the dark sides of the force then space-time, the arena where everything we know takes place, could be bent and twisted with infinitely greater ease than was ever suspected. This would open Pandora's box to everything from vacuum energy weapons of mass destruction (capable of destroying the universe!) to space-time warp drives and time machines."

Exactly, the above is the thesis of all my books since 2002 at least.

"A quick survey of the international electronic archive of physics papers at www.arXiv.org shows that research into the vacuum of space-time for energy production is alive and well. Most authors are independent researchers struggling with limited funding and resources, yet their theoretical results suggest that somewhere in Nick Cook's black world, a major breakthrough has already taken place. Most likely Israel, UK, the United States and Russia are in the lead, but China, France, Serbia, Ukraine, Iran, India and Saudi Arabia all have scientists actively pursuing the fundamental physics that determine the fabric of our reality, and are seeking the theory and the means to access the enormous energies locked inside of the vacuum since the creation of the universe. Even if the black budget world has yet to unleash the enormous potential of vacuum energy, there are signs that those in power may have begun to take notice. Dr. Harold Puthoff, a scientist with strong government connections, who has previously worked on classified projects for the CIA, is a major proponent of vacuum energy physics. Nick Cook's book, "The Hunt for Zero Point", and his recent stories on zero point energy in "Jane's Defence Weekly" have also brought attention to the dangers and military potential of exotic vacuum research. The American intelligence community financed so-called psychic spies for over twenty years and through four presidential administrations. It is highly unlikely that they would ignore the potential of the quantum vacuum. Dr. George Chapline, of the Lawrence Livermore National Laboratory, and Dr. Jack Sarfatti in San Francisco, knew each other in the sixties in La Jolla, have independently proposed that the quantum vacuum may unstable to the formation of coherent virtual processes. Sarfatti suggests that gravity is an emergent property determined by the physics of the vacuum. His idea is to find a means of directly interacting with the physics of the vacuum that controls the shape of space-time. Such a possibility would be consistent with the reported success of Evgeny Podkletnov, the Russian scientist who is experimenting with spinning superconducting disks. Podkletnov's most recent papers report the appearance of a mysterious coherent beam of "gravity like radiation" with a measured force of 1000 G. In an interview on BBC radio, Nick Cook pointed out one immediate application of the Podkletnov beam - the destruction of missiles and satellites in flight or in orbit around the earth. Cook showed the BBC internal documents from Boeing, the American aerospace contractor, proving their interest in Podkletnov's research."

This beam stuff I am suspicious of. Of course if the experiment is good, I have to think more about it. I am not so sure if Podkletnov's experiment is any good and has been replicated.

"The connections between Podkletnov's results, and the kind of vacuum research explored by Sarfatti, beginning in 1999 at the International Space Sciences Organization are the latest threads in a trail that most likely originates in cold war disinformation, a game played by East and West against each other. Glasnost has shifted the balance of partnerships and the positions of the players, but not the stakes of an outcome that would leave the world with even more prolific and powerful weapons of mass destruction."

That is true, as shown in "Destiny Matrix," however you leave out the most important evidence -UFOs!

"When I first contacted Nick Cook, he wrote "What intrigues me in this whole business are the connections." Although Nick Cook never revealed the identity of his "deep throat" contact called "Dr. Dan Marckus" in the book "The Hunt for Zero Point", there was no question that the Podkletnov results had played a major part in fitting together the pieces of the puzzle. The amount of interest was in Podkletnov's reports by NASA, Boeing, and others in the international arena of aerospace and military research communities was evidence that there was more here to explore than the latest musings of the intellectual elite. The truth is that a fundamental theory of gravity at the scales of subatomic nuclear physics does not exist. The fact is that no one understands the nature of the gravitational field at very small scales. In fact gravity has barely been probed much below one millimeter. Every attempt to unify the physical theories of gravity with the well-known standard model physics of electromagnetism, and the strong and weak nuclear forces, has failed. More importantly there has been recent progress in the exotic areas of mainstream research, such as superstring theory, which suggest new kinds of physics, which might support explanations for Podkletnov's "impulse gravity" effect. One of the current fads in theoretical physics involves large extra dimensions of space that allow a much stronger version of gravity to leak off the membrane world of our ordinary three dimensions. The large dimensional picture allows for the well known forces of electromagnetism, and the strong and weak nuclear forces, to be confined to a three dimensional "brane-world" floating in a higher dimensional hyperspace. Gravitons, the quanta of the gravitational force, are viewed as closed strings, and are able to slip off of our "brane-world", which explains why the gravitational force is so much weaker than the other forces that hold matter together. Gravitons could be exchanged between our brane-world and another brane floating nearby in the same higher dimensions.
The Sarfatti picture offers a more direct interaction with the new physics than the brane world ideas. Sarfatti's vision is to find a means of using electromagnetic fields in the Josephson effect to couple to the virtual electron-positron pair giant coherent condensate inflation field inside the vacuum that controls the shape of space-time to the real electron pair giant coherent condensate of a control high temperature superconductor. UCB's Ray Chiao has a similar idea using a superconductor to transduce electromagnetic far field waves to gravity waves with high efficiency

conversion. Sarfatti wants to do the same thing with non-propagating electromagnetic and gravity near fields. One wonders if the black budget world may have already produced some of the technology needed to explore and test these new realms."

Hal Puthoff wrote:

"Those with a practical bent of mind may be left with yet one more unanswered question. Can this emerging Rosetta Stone of physics be used to translate such lofty insights into mundane application? Could the engineer of the future specialize in "vacuum engineering?" Could the energy crisis be solved by harnessing the energies of the zero-point sea? After all, since the basic zero-point energy form is highly random in nature, and tending towards self-cancellation, if a way could be found to bring order out of chaos, the, because of the highly energetic nature of the vacuum fluctuations, relatively large effects could in principle be produced. Given our relative ignorance at this point, we must fall back on a quote given by Podolny (12) when contemplating this same issue."

Hal Puthoff's remark above that the randomness of the zero point energy tends toward self-cancellation is suspect. The mean is zero, but the root mean square fluctuation is not zero. Neither Hal Puthoff nor Bernie Haisch have ever explicitly mentioned "vacuum coherence" in any of their published papers nor have they mentioned the role of w = pressure/energy density = -1 for the zero point energy, nor have they explicitly included the virtual electron-positron pairs in spite of Hal's use of "PV" for "Polarized Vacuum" in their allegations that the rest mass of the electron etc and gravity itself emerge from only the transverse electromagnetic zero point fluctuations that they interpret semi-classically as "stochastic electrodynamics" with h as some kind of fudge factor. Excerpts from Eric Davis's MUFON 2001 paper. What he says in other parts of the paper about how Puthoff's theory can explain this data makes no sense to me at all, are extremely misleading to Pentagon, CIA et-al who allegedly read them, and, for the record, I strongly disagree with his assessment there on how to properly understand the UFO "facts" listed below. This list is based on a list made by NIDS in Las Vegas, which funded by multi-millionaire Robert Bigelow[lxxvi], employed a significant number of former US Military, Intelligence, FBI and police personnel. Davis's list below is not the work of wild-eyed kooks and crackpots. Therefore, knee-jerk debunking aside, the purely empirical allegations Davis makes below, and he is a Ph.D. physicist, are obviously of extreme military importance if true.

UFO Phenomenology

Paul Hill has already delineated and characterized UFO performances and dynamics in his excellent book. From a rigorous aeronautical and physics analysis of many cases (the unexplainable, non-prosaic ones), Hill concluded that UFOs are craft that would have to utilize an engineered "acceleration-field" technology in order to manifest their various performance characteristics. "Acceleration-field" is the old fashioned term for space-time metric. Wormhole-stargates and the Alcubierre warp drive metric are examples of modern space-time metric engineering concepts, both of which require engineering of the vacuum to mine the negative energies needed to generate such metric modifications."

Eric Davis makes the common error of confusing the required strong "negative pressure" with negative energy density. In fact you want positive zero point energy density with its equally strong negative pressure.

"Jacques Vallee has also analyzed UFO cases over four decades and summarized his findings in several excellent, groundbreaking books and articles.... Vallee concluded that UFO phenomenon is consistent with a technology centered on a craft using a very revolutionary propulsion system, which possesses an anti-physical dimension in addition to others."

I introduced Jacques Vallee to Francis Ford Coppola and his friends in 1976, which is how Vallee got to work on Stephen Spielberg's "Close Encounters of the Third Kind."

Saul-Paul Sirag, who was there, tells this story in my book "Destiny Matrix." Vallee is wrong about the "anti-physical dimension" as this book shows in detail. Physics has advanced way beyond Vallee's understanding since he made that unfortunate remark.

"The phenomenon is the product of a technology in the sense that it is a real, physical, material object. The physical characteristics of UFOs is as follows ...
- *witnesses describe an object that occupies a position in space*
- *moves as time passes*
- *interacts with the environment through thermal effects as well as light absorption and emission*
- *produces turbulence*
- *when landed, leaves indentations and burns from which approximate mass/energy figures can be derived*
- *gives rise to photographic images*
- *gives rise to electric, magnetic and gravitational disturbances*
- *But UFOs also manifest anti-physical effects by using advanced physical principles. These anti-physical effects are as follows..."*

I object to the use of "anti-physical" in this context. UFOs, if real and not pure bunkum propagated by evildoers and only believed by gullible fools as "skeptics" say, are physical phenomena as are the alleged paranormal influences on our conscious minds. The set of "anti-physical" phenomena is the empty set.

- *sinking into the ground*
- *shrink in size, grow larger, or change shape*
- *becoming fuzzy and transparent on the spot*
- *divide into two or more craft, several of them merge into one object at*
- *disappearing at one point and appearing elsewhere instantaneously*
- *remaining observable visually while not detected by radar*
- *missing time/time dilation*
- *topological inversion/space dilation (UFO was estimated to be of small exterior size/volume, but witness(s) saw a huge interior many times the exterior size)*
- *balls of colored, intensely bright light under intelligent control*
- *Doppler blueshifting and redshifting effects of moving and motionless UFOs"*

The zero point energy density of exotic vacuum will cause such frequency shifts. Indeed, an Unconventional Flying Object using zero point pressure propulsion will show a universal "Reverse Doppler Effect" for all signals including sound waves in air, as alleged by Bruce Cornet in http://bcornet.homestead.com/files/weirdsound.htm . However, I am not endorsing the authenticity of Cornet's allegations. I leave that to experimentalists like Bruce Maccabee to evaluate. Negative quantum zero point pressure dark energy causes a gravity blue shift whilst positive quantum zero point pressure dark matter causes a gravity blue shift. The strength of such effects is not limited by the extreme weakness of Newton's gravitational constant G, which is sidestepped. Imagine a UFO coming straight at you. The positive quantum zero point pressure dark matter boundary layer at its bow will cause a gravity warp red shift tending to cancel the motional Doppler blue shift. As the UFO recedes from you, the negative quantum zero point pressure dark energy boundary layer will cause a gravity warp blue shift tending to cancel the motional Doppler red shift. Indeed, mapping the frequency shifts of radar signals deflected from all parts of the UFO will reveal much about the exotic vacuum field configuration of its weightless warp drive in which there are no g-forces on the objects inside the presumably alien machine that may well have time travel capability.

"The physiological reactions caused by UFOs are ...
- *Burns*
- *sounds (beeping, buzzing, humming, sharp/piercing whistling, swooshing/air rushing,loud/deafening roaring, sound of a storm, etc.)*
- *vibrations*
- *partial paralysis*
- *extreme heat or cold sensation*
- *odors (powerful, sweet or strange fragrance, rotten eggs, sulphurous, pungent, stinking, muskylike,etc.)*
- *metallic taste*
- *pricklings*
- *temporary blindness when exposed to the object's light*
- *nausea*
- *bloody nose and/or ears; severe headache*

- *difficulty in breathing*
- *loss of volition*
- *drowsiness in the days following a close encounter*
- *There are psychic effects triggered by UFOs either purposely or as a side effect of the presence of the UFO. These are ...*
- *impressions of communication w/o direct sensory channel*
- *levitation of the witness or of objects and animals in the vicinity*
- *poltergeist phenomena: motions and sounds w/o a specific cause, outside of the observed presence of a UFO*
- *maneuvers of a UFO appearing to anticipate the witness' thoughts*
- *premonitory dreams or visions*
- *personality changes promoting unusual abilities in the witness*
- *healing*

Wormhole-stargate characteristics can fulfill Hill's and Vallee's criteria for UFO phenomenon in both the physical and anti-physical sense. ... The items marked with single asterisks are explainable as manifestations of a wormhole opening up and intersecting our local space, the negative energy used"

There is no mention of "dark energy" in Eric Davis's paper nor any mention of it in any of Hal Puthoff's metric engineering papers. I am most likely the first physicist with valid credentials to make this connection. Eric's use of "negative energy" is too vague and should be replaced by "negative pressure" for a w = -1 quantum zero point dark energy exotic vacuum field with the required anti-gravity of sufficient strength for practical metric engineering super-technology that, if achieve, would render all of our conventional military weaponry impotent and obsolete. The evidence Davis presents is that someone out there already has this capability making us all look like the Lilliputians in Jonathan Swift's "Gulliver's Travels."

Big Foot visits Robert Bigelow's Ranch from The Universe Next Door?

Note added in this second edition. The book "The Hunt for Skinwalker" by Colm Kelleher and George Knapp appeared. Colm was scientific director at NIDS and this book has many important facts showing that we are, in a low-intensity conflict with some kind of advanced alien intelligence. More data on this is in the new book "Exempt from Disclosure" by former USAF Intelligence agents Robert M. Collins and Richard C. Doty. They claim the involvement of CIA's James Jesus Angleton and other key historical USG figures. Collins and Doty with Tim Cooper claim living ETs were with USG personnel. They also claim an alien energy device the size of a laptop that can output 100,000 watts at 100 amps continuous that never runs out – not in last 60 years at least. Linda Moulton Howe on George Noory's C2C radio show January 2006 claims a shooting war with ETs started in 1947 with scores of US planes shot down by saucers in a few months and V2 tests in New Mexico being attacked. She also claimed on the February 10, 2006 C2C show that the strongly telepathic "EBENS" (ETs) demonstrated their psychokinetic power to the US Military by lifting 200 lb metal chunks with their minds. All of these alleged powers, if true, show "signal nonlocality" in strong violation of the basic postulates of quantum cryptography. Scott Littleton, retired anthropologist from Claremont Colleges, is writing a book about 1942 saucer attack on Los Angeles in 1942 that he actually witnessed as a small boy. More details can be found on Google.

Metric engineering with the digitized loop quantum gravity string, area and volume operators of spin foams reduces space-time physics to the quantum computing of virtual realities like in The Matrix films. Eric Davis in MUFON 2001 reports that NIDS saw a "Big Foot" Sasquatch figure pop out of an artificial wormhole from the universe next door. This is not distant in concept from what Fredkin, Tipler and Wolfram have proposed except we need qubits not c-bits. The hologram picture and the Bekenstein bits on the black hole horizon also suggest this.

"... to generate the wormhole, and/or of effects from the scout craft/probes that came through the wormhole. Items marked with double asterisks are explainable as reactions triggered by the scout craft/probes that come through the wormhole. Items marked with a dagger can only be explained as consciousness-related effects that are triggered on purpose or as a side effect of the presence of the intelligence controlling the wormhole and scout craft/probes. This goes to the heart of the incommensurability or cognitive mismatch problem that exists between the human race and the intelligence responsible for UFO phenomenon ... Wormhole and other UFO-related manifestations can be instrumented in the field with a sensor array system to measure and record their physical effects. Stride (35) developed a proposal for a passive autonomous data acquisition platform using COTS hardware to collect sensor data on phenomenon. NIDS deploys a subset of such portable sensor devices in the field. Germane to the wormhole issue are sensors that measure acceleration/gravimetric effects, electric and magnetic effects (AC/DC), spectroradiometry, temperature/pressure changes, optical and acoustic radiometry, radio/microwave, visible/IR/UV imaging, and so on. Miniaturization of electronic sensor instruments has allowed such devices to become very small, lightweight, portable and easily integrated into reasonably sized packages for field

deployment. These items are detailed in (35) and elsewhere in the technical literature.

A wormhole-stargate has never been constructed in the laboratory, so a full accounting of its physical effects is not yet known beyond what is obvious from recent theory. This also applies to the generation of large amounts of negative energy, which will likely, if it is possible, to have a technology with byproduct effects that are not determinable at this time. Laboratory experiments on the vacuum using ultrahigh intensity lasers will be necessary to ascertain these issues."

I disagree. Using these lasers is not relevant to the UFO technology in my opinion.

"3.1 Example Cases from the NIDS Database

UFO witness descriptions are the database presently available for examining the wormhole hypothesis along with the meager physical data acquired by surveillance equipment (see for example, references 36-39). And we recognize that witness reports are not rigorous from the standpoint of collecting physics data. Of the more than 650 cases investigated by NIDS, several dozen clearly portend wormhole manifestations. Particular examples include field research NIDS conducted in northeastern Utah whereby the following example data was acquired:

- *intensely bright, colored balls of light under intelligent control; either monochromatic or changing color; possessing either smooth or variable liquid turbulence-like surface/internal texture; maneuvering/hovering near people and around property; brightening or fading and blue/red Doppler shifting when appearing or disappearing*
- *very large, very bright orange-colored opening in the daytime sky; a completely different or foreign looking sky was seen through the opening; an object was seen (through rifle spotting scope) moving through the opening at rapid speed*
- *faint light appears in the air a few feet above a dirt road; light grows in intensity becoming very bright; bright light then becomes a hole that opens up (growing from 1 to 3 feet diameter) and from within which another light is emanating; a large, black creature (~ 400 lbs., 8 to 9 feet tall) is seen crawling out of the hole (as seen through 3rd generation military night vision, hole appeared 3-dimensional with tunnel-like interior), it stood up and ran away into the surrounding dark of night; the brightly lit hole closed and faded away"*

Big Foot? See the book"Hunt for the Skinwalker by Colm A. Kelleher and George Knapp for more details about the strange Utah Ranch.

ISSO Torsion Field Physics Workshop 2000

America's leading physics expert in torsion field physics is Richard Hammond at North Dakota State College in Fargo. Richard visited us at the ISSO Torsion Workshop in 2000 that included J.P. Vigier, Bernie Haisch, Robert Kiehn, Giovanni Modanese, Gennady Shipov, Creon Levit, Saul-Paul Sirag, Vladimir Poponin and other physicists from CIPA and elsewhere. Hammond does both theory and experiment. He reported some positive results on torsion wave propagation at ISSO but also told us that the US Navy cancelled his contract when he started getting results. Akimov in Russia of course alleges that torsion waves can be used for secure submarine command control communication. We can only speculate that the US Navy took Hammond's project into the "Black World." The reader can find reliable facts about torsion field physics from Hammond's Reports in Progress in Physics paper from 2002. Our torsion field study started at ISSO 1999-2000 when several million dollars was spent trying to reverse engineer the alien ET technology for the USG Intelligence Community. We did not have direct access to the alleged alien ET saucers in captivity, but we were "told" about them. Our Russian comrades were also working on this and we invited them to San Francisco several times along with J.P.Vigier and other physicists. The attached is the theory that several million dollars paid for. The experiments at SARA in Huntington Beach that took most of them money did not pan out. RP at Langley was kept informed about all this at the time. Therefore, Neo Con polemicist Stephen Schwartz's allegation that we were violating the US Serbian embargo has no merit. I personally was not even involved and never met the Serbs Maric and Dragic, nor communicated with them. The whole operation was handled by Creon Levit. Schwartz also falsely accused celebrity Michael Savage[lxxvii] of The Savage Nation Radio Talk Show with millions of listeners of "working for the Serbs." Schwartz, Savage and I all know each other from North Beach, San Francisco since the late 1970's and Schwartz, a self-proclaimed "Red Diaper Baby – Trotskyite" convert to Islam, has always felt intensely competitive with Savage who essentially ignores him. We were simply checking to see what progress, if any, the Serbs were making in this new possible form of atomic energy that might be related to cold fusion. Frank Tipler, coauthor of the monumental "The Anthropic Principle" with John Barrow was reported by John Brockman at Edge.Org in early March 2006 that he was worried that terrorists could convert 100 kg of matter *completely* into energy using simple technology in a garage workshop. This would be consistent with Sir Martin Rees's worries in "Our Final Hour." Note, in contrast, that a nuclear bomb only converts a small fraction of its mass into energy.

Topics In Cutting Edge Physics Today

Just Six Numbers

What are the six numbers that control the Universe we live in? The idea is that there are an infinity of infinity of parallel universes limited by the Hubble horizons of their past light cones from any observation point that reach back to the Big Bangs creating each one of them in quantum vacuum phase transitions that break symmetries. The Weak Anthropic Principle (WAP) of generalized Darwinian natural selection applies. There is no mystery to the "fine-tuning" of these 6 control parameters. Intelligent Design is not needed. God is not needed -- at least not for this aspect of The Problem. The Charismatic Chain of Hermetic Adepts, which, perhaps, included Isaac Newton, has been working on The Problem for at least a thousand years starting in Troyes, France – or so The Legend of The Da Vinci Code and Henry Miller on Carlo Suares alleges.

Dark Energy Warp Drive

On Jan 29, 2004, at 12:02 PM
On UFOS, "metric engineering" and the recent amazing completely unexpected discovery of Dark Energy/Matter as 96% of the Universe.

"Who ordered that?" Isidor Isaac Rabi

"The Question is: What is The Question?" John Archibald Wheeler

"Why not go to a lower vacuum here and now? ... It was a very frightening experience which I had when I first thought about these bubbles ... I really shivered ... I told Andrei Sakharov about these bubbles. I vividly remember his reaction. He said: 'Such theoretical work should be forbidden. It's too dangerous." Lev. B. Okun hep-ph/0112032

1. "Dark energy" is almost 2/3 of the Universe and is a repelling anti-gravity "acceleration field" in the precise sense of Paul Hill's good book "Unconventional Flying Objects."

2. "Dark matter is almost 1/3 of the Universe and is an attractive gravity field that is not made of the elementary particles the stars and us are made of.

3. Ordinary matter and light is only a very tiny part of the Universe.

4. Put a chunk of dark energy near a chunk of dark matter and you basically have a weightless Alcubierre warp drive that Robert Forward called "negative matter propulsion" and that Paul Hill in his excellent book "Unconventional Flying Objects" called "acceleration field" propulsion. This means you feel weightless in free float like the astronauts in space with their rocket engines switched off and the Universe passes by you seemingly faster than the speed of light. Herman Bondi, a consultant to British Intelligence, had a glimpse of this idea forty-five years ago before dark energy was discovered. I attended his Cornell lecture on the subject back then. So did Joseph Stalin's physics spymaster identified by Herbert Romerstein et-al the Venona Project Yacov Terletskii. You can time travel to your past and to your future and beyond under certain conditions. Metric engineering is the control of exotic vacuum coherence by electromagnetic tuning of its holographic phase. The usual stress-energy density tensor of matter and electromagnetic field (AKA "Mass-Energy) T_{uv}, that Hal Puthoff relies upon in all his "action" calculations, is not at all relevant to the problem because space-time is too stiff to bend for practical free float weightless propulsion through the time travel Star Gates in Carl Sagan's "Contact" that Kip Thorne helped him with. All of these "sci fi" ideas are now possible in fact because of the 1998 discovery of anti-gravitating geometrodynamic "dark energy." We are on the brink of making "Star Trek" real including the "Men Like Gods" of the "Q-Continuum." It is curious that Gene Roddenberry, the creator of "Star Trek" was, for awhile part of Andrija Puharich's paranormal UFO study group around the time of the Uri Geller phenomenon described by

Martin Gardner in "Magic and Paraphysics" in "Science: Good, Bad and Bogus."

5. Einstein's theory of special relativity from 1905 is the unification of space with time into a rigid "space-time" and the unification of matter and energy.

6. Einstein's general theory of relativity shows that space-time is not rigid but is warped not only by energy but also even by itself in a self-organizing way.

7. Quantum theory is about the informational relationship between the observer and the observed. Some important physicists like John Von Neumann, Eugene Wigner and Roger Penrose thought that our inner consciousness plays a key role in quantum physics. Other important physicists violently disagree. It depends what you mean by "quantum theory."

There are many different ways of interpreting quantum theory and its boundary of validity is not yet understood especially in the strong gravity field of a black hole. A key issue is "signal nonlocality," i.e. the use of "entanglement" as a direct communication channel without an auxiliary light speed limited classical channel as in quantum teleportation and cryptography where Eve cannot tap the channel between Alice and Bob without them knowing it. These applications require irreducible quantum randomness that is thought to forbid "signal nonlocality," i.e. to enforce "signal locality" or "passion at a distance" (Abner Shimony). The analysis of my old Cornell chum Lenny Susskind, for example, on information loss behind the event horizon of a black hole, depends on "signal locality" as does the non-Boolean "topos""quantum logic" of cosmology mentioned by Lee Smolin. Yet, the Bohmian analysis of Antony Valentini at Cambridge University and the International Centre for Theoretical Physics in Trieste, Italy suggests otherwise. The idea here is that orthodox nonlocal micro-quantum theory with irreducible randomness and consequent signal locality is only the "sub-quantal" (in sense of Jean Pierre Vigier) thermal equilibrium limit of a more general non-equilibrium local MACRO-Quantum theory for cosmology and the emergence of "classical curved space-time" with signal nonlocality.

8. The cutting edge of physics today in 2004 is "quantum gravity", which has two main schools of thought, string theory and loop theory. The two theories may simply be dual dimensional images of each other. The basic object of loop theory is a two-dimensional undivided quantum of "area" or Shannon "c-bit" and the basic object of string theory is a one-dimensional undivided quantum of "length" or "qubit", which together make a three-dimensional "space." You can think of the "string" simultaneously as Brian Greene's one-dimensional "vibrations of pure energy" in a space of ten dimensions and as the strings in a quantum computer program that is beyond space-time. Loop theory describes three-dimensional "space" as zero-dimensional "point" on a kind of quantum computing "spin network." When you add a kind of proto "time" to it you get a "spin foam." Two-dimensional area is described by a 1-dimensional stringy "edge" on the spin network. The duality is intuitively obvious. The 1-dimensional physical string is really like a linked chain with each link as a "bit" of information not unlike the purely mathematical "strings" of computer theory. No matter that the string is embedded in a higher dimensional "hyperspace" since all but three of those extra "spacelike" dimensions

are perhaps curled up into tiny circles with radii called "moduli" forming what the mathematicians like to call a "Calabi-Yau space." On the other hand, the parallel "brane worlds" model does not compactify these extra dimensions into little multi-dimensional tori at all. Brian Greene calls all these wild speculations by the physics establishment ingroup "The Elegant Universe" in which the spin foam is "weaved" into the fabric of reality. God is seen by Brian as a kind of tailor or fashion designer reminding me of the Vatican scene in Fellini's "Roma."

9. Both the string and the loop theorists think that the quanta of length and area respectively are fixed constants that are always very tiny. If UFOs are fact and not fiction, then they are, most likely, badly mistaken in that belief.

Then again, I may be mistaken.

10. Quantum wholes are greater than the simple sum of their parts. The quantum information comes in two forms called "active" and "inactive." This is explained in detail in David Bohm's and Basil Hiley's "The Undivided Universe." Entanglement is important in quantum computing applications like "teleportation of qubits" and "untappable cryptography." It also is the key idea in "environmental decoherence", which is an attempt, only partly successful, that tries to explain the irreversible thermodynamic flow or "arrow of time" and the "collapse of the quantum state" or "Von Neumann projection", i.e. why the large-scale world of ordinary experience seems "definite" without us being alive and dead at the same time in the same place as a naïve extrapolation of the quantum properties of tiny simple objects suggests. Note that while our inner perceptions of the outer world seem definite without the same object being in two places at the same time, our pure inner conscious thought has quantum superposition in which we hold two, or more, incompatible ideas in our mind's eye simultaneously as in Hamlet's speech "To be, or not to be. That is The Question …" On the other hand John Archibald Wheeler says "The Question is: What is The Question." "Quantum logic" is the study of "quantum binary questions" or qubits arranged in a non-Boolean partially ordered lattice. This is very different from our computers whose logic is that of a Boolean lattice. The non-Boolean lattice, where each question is a node on a graph, has "Isles of Boolean lattices" corresponding to compatible questions that can be answered definitely simultaneously with the same configuration of detectors. There is an approach to quantum gravity called "consistent histories" which consists of a "story line" of questions (Lee Smolin's "The Three Roads to Quantum Gravity") the problem is that knowing which questions to ask to tell the story is an insolvable problem, or perhaps I should say, is an undecidable question in the sense of Godel's "incompleteness theorem" of 1931 and Its corollaries like the "Halting Problem" of computer theory asking the question "When exactly will the program stop?" So this is what Wheeler is alluding to in his cryptographic remark: "The Question is: What is The Question?"

Now first thing we need to understand in looking at Rovelli's version of Lee Smolin's "Three Roads to Quantum Gravity" is that all three roads are top -> down. But there is a Gurdjieffian "Fourth (bottom -> up) Way" of "emergence", ignored completely by Smolin and Rovelli and All The King's Men trying to put Humpty Dumpty together

again, due to the great Soviet physicist Andrei Sakharov and also the Princeton physicist P.W. Anderson. Sakharov called it "metric elasticity." Anderson called it "More is different" with an information-rich (low thermodynamic entropy) giant quantum coherence field that is local without entanglement with "generalized phase rigidity" making it immune to "environmental decoherence" that in a special case is Sakharov's metric elasticity and also the "tension" of "string theory." The fact that the giant quantum coherence field is local automatically explains why the outer world is definite without the same object being in two places at the same time and why we are not alive and dead at the same time like Schrodinger's Cat out of a story by Lewis Carroll. As Wheeler wrote: "Physics is simple when it is local." P.W. Anderson's "More is different" explains why large –scale physics is local without needing Oxford's David Deutsch's "excess metaphysical baggage" (Wheeler's term) of the quantum "Multiverse" of "shadow objects." This is not to exclude "parallel universes", but one must be vary careful on how "universe" is defined. It all depends on what you mean by "is." It's the ontology stupid!
lxxviii

O Brane New Worlds

"O brave new world that has such people in't"
(Miranda, from Shakespeare's 'The Tempest', Act V)

George Dvali in February, 2004 Scientific American wrote:
"When gravity operates over microscopic distances —for instance, at the center of a black hole, where a huge mass is packed into a subatomic volume—the bizarre quantum properties of matter come into play, and string theory describes how the law of gravity changes. Over greater distances, string theorists have generally assumed that quantum effects are unimportant. Yet the cosmological discoveries of the past several years have encouraged researchers to reconsider. Four years ago my colleagues and I asked whether string theory would change the law of gravity not just on the smallest scales but also on the largest ones. The feature of string theory that could bring about this revision is its extra dimensions—additional directions in which particles can roam. The theory adds six or seven dimensions to the usual three."

Theorists today are willing to pay any price to avoid signal nonlocality – even extra dimensions. Well, perhaps, that is not too high a price after all?

"In the past, string theorists have argued that the extra dimensions are too small for us to see or move in. But recent progress reveals that some or all of the new dimensions could actually be infinite in size. They are hidden from view not because they are small but because the particles that make up our bodies are trapped in three dimensions. The one particle that eludes confinement is the particle that transmits the force of gravity, and as a result, the law of gravity changes."

There is one nice feature about this idea of enormous "branes" in that it easily explains why there is no significant amount of antimatter in our universe. According to John Archibald Wheeler's "Geometrodynamics" from the 1950's, the elementary particles, that today we know as the "lepto-quarks" are in reality David Bohm's "hidden" or "extra variable" that are tiny wormholes in three dimensional space with quantized gauge force fluxes threading them like we see in Type II superconducting "vortices" in condensed matter physics near absolute zero temperature. Wheeler called this "Mass without mass" with "Charge without charge" and "Spin without spin." Wheeler's idea did not work back then almost fifty years ago because gravity was thought to be too weak. In other words, space-time geometry was simply too stiff to bend easily. Space-time's "string tension" was too large. This implied that Wheeler's wormholes or "geons" were simply much too big to be the elementary particles like the electron. Indeed, this is just like the speed of sound barrier was mistakenly thought to be before the era of jet planes and Chuck Yeager's breakthrough. However, today in 2004 we know that it is quite plausible that Andrei Sakharov's "metric elasticity" for emergent gravity out of zero point vacuum energy fluctuations, or what Ed Witten, (The Big Cheese of "M-Theory") calls alpha' can be a scale-dependent variable not an immutable constant. Witten's alpha' is proportional to the reciprocal of the string tension, that originated in the observed universal "gravity" "Regge parallel trajectories slope" of the "hadronic resonances" of $(1 \text{ Gev})^{-2}$ unceremoniously kicked up to the quantum gravity scale of $(10^{19} \text{ Gev})^{-2}$ is a variable that gets less stiff at smaller scales of higher energy transfer in scattering experiments. Indeed, the effective strong short-range gravity, suggested by Abdus Salam in the early 1970's, solves more than one hitherto unsolved fundamental conceptual problems in high-energy physics. Strong gravity at the scale of 1 fermi $(10^{-13}$ cm) is 40 powers of ten stronger than Newton's gravity. This means Wheeler's wormholes are just the right size. It's like those size-changing drugs that Alice takes in Lewis Carroll's Wonderland. First of all it solves the mystery of the missing anti-matter called "broken charge conjugation symmetry" or "C-violation." The missing antimatter is in a parallel "brane world" next door across a thin barrier of extra dimensional "hyperspace." Imagine a wormhole with two mouths or ends attached to two different brane worlds with infinitely large uncompactified extra space dimension as Dvali explains. The Faraday flux lines of electro-weak-strong gauge forces must form closed loops. Flux lines coming out of one mouth of the wormhole must enter the other mouth of the same wormhole. We know from Gauss's divergence theorem that the sign of the effective charge changes from + to − depending on whether the flux lines leave or enter the wormhole mouths that you can picture as little 2-dimensional spherical surfaces in ordinary 3-dimensional space. These spheres may themselves not be simply connected but may have little wormhole handles on them as well. It may be Ezekial's vision of wormholes on wormholes. This might mean only a single electromagnetic flux in which the weak and strong forces are simply topological variations of the same theme. That would be a nice unification hardly less speculative than what you see on Brian Greene's "Elegant Universe" on NOVA. One famous Dead Physicist Ludwig Boltzmann said "Elegance is for tailors." If NOVA was produced by BBC they might call it "Saville Row" physics – very fashionable. Therefore, there are equal amounts of matter and anti-matter but the corresponding wormhole mouths with equal and opposite quantized flux charges are attached to different Level I parallel Hubble sphere universes next door to each otheron the same Level II inflation bubble! All the trapped fluxes point the same way in a single Hubble sphere "domain" analogous to the domains of a ferromagnet. Indeed it's more like an anti-ferromagnet with the hadronic charges pointing opposite to the leptonic charges in a single Hubble sphere local universe with its past light cone causal horizon at every point event in it. There is still another problem solved here. It is the problem that Richard Feynman tried and failed and All The King's Men have not solved it to this day. The problem is the infinite self-energy of the point electron. Why a point electron you ask? Because, there was, hitherto, no way to prevent the extended electric charge distribution of a non-point electron from exploding! This unsolved problem is more than one hundred years old! It's solved now. The Lorentz-Abraham stress is simply the strongly short-range attractive exotic vacuum "dark matter" zero point fluctuation "positive pressure" from negative zero point energy density where the throats of Wheeler's wormhole "geons" are like the vortex cores of superfluids where the vacuum coherence field drops to zero on the scale of a "coherence length." Note that the trapped "near field" quantized Faraday flux lines of virtual spin 1 quanta in coherent states are contained in the Meissner effect's "penetration depth" that is no more than one over the square root of two of the coherence length like in a Type II superconductor.

"WHEN ASTRONOMERS ENCOUNTERED the cosmic acceleration, their first reaction was to attribute it to the so-called cosmological constant. Notoriously introduced and then retracted by Einstein, the constant represents the energy inherent in space

Maybe cosmic acceleration isn't caused by dark energy after all but by an inexorable leakage of gravity out of our world."

"Cosmologists and particle physicists have seldom felt so confused. Although our standard model of cosmology has been confirmed by recent observations, it still has a gaping hole: nobody knows why the expansion of the universe is accelerating."

Warped Passages: strong short-range hyperspace brane gravity

Note that the brane world theories have a larger gravity force at small distances. The brane world theory keeps all quanta stuck to the 3D brane except for gravitons that escape into the extra dimensions. This is unlike the Kaluza-Klein theory of extra dimensions where all the quanta can escape to the extra dimensions. The theory goes like this: The Kaluza-Klein extra dimensions are curled up to a size $R < (10^3 \text{ Gev})^{-1}$ from collider experimental data. The brane extra dimensions, in contrast, can be much larger, indeed infinite *if the brane space is curved* as shown by Lisa Randall (with Sundrum). The compact extra brane dimensions however cannot be larger than a fraction of a millimeter based on Newton's law data up to the beginning of 2006. The 4D Planck mass M_{Planck} corresponding to 10^{-33} cm is a projection from the brane hyperspace of $4 + n$ dimensions where the n extra dimensions are curled up in a hyper-torus $S^1 \times ... S^1_n$ on the scale R

$$M^2_{Planck} = M^{2+n}_{Hyperspace} \left(\frac{cR}{h} \right)^n$$

$$\left(\frac{M_{Planck}}{M_{Hyperspace}} \right)^2 = \left(\frac{M_{Hyperspace} cR}{h} \right)^n = \left(\frac{R}{\lambda_{Hyperspace}} \right)^n$$

$$\frac{G_{Hyperspace}}{G_{Planck}} = \left(\frac{M_{Hyperspace} cR}{h} \right)^n = \left(\frac{M_{Hyperspace}}{m_{Hyper\_Graviton}} \right)^n$$

Where $\lambda_{Hyperspace}$ is a quantum Compton wavelength scale that is the effective Planck length in brane hyperspace. From Gauss's flux law, weak Newtonian gravity is found when $r > R$, but the effective gravity is strong when $r < R$. Note that the effective strength of gravity is

$$G \sim \frac{hc}{M^2}$$

The brane theory is equivalent to the old Abdus Salam bi-metric f-meson theory where the spin 2 f-meson has an effective mass

$$m_f = \frac{h}{Rc}$$

with strength

$$G_{Hyperspace} = \frac{hc}{M^2_{Hyperspace}}$$

with the effective static Yukawa potential energy

$$U_{Hyperspace} = -\frac{G_{Hyperspace} mm'}{r} e^{-\frac{r}{R}}$$

We want

$$\frac{G_{Hyperspace}}{G_{Planck}} ? \; 1$$

$$R \sim 10^{-33} cm$$

The extra dimensions can be infinite with an effective compactification from the bulk curvature.

The Randall-Sundrum Model

Assume 4 + 1 space-time rather than the 3 + 1 we appear to live in. Assume a negative dark matter cosmological constant i.e. negative energy density with positive pressure and w = -1.

$$\Lambda = -\frac{1}{l^2}$$

The symbol l should not be confused with the curved tetrad elsewhere in this book using the same symbol.

The 5D Einstein field equation is[lxxix]

$$G_{AB} + \Lambda g_{AB} = \kappa^2 T_{AB}$$

$$\kappa^2 = \frac{1}{M_5^3}$$

$$n = 1$$

$$M_{Hyperspace}^{2+n} \equiv M_{4+n}^{2+n}$$

Hyperspace is empty in brane theory because the open strings that are the leptons, quarks, photons, gluons and weakons have their ends stuck to the brane like flies on flypaper. Therefore, only the bulk curvature of hyperspace contributes. This bulk curvature comes from the vacuum energy $\sim \Lambda$. The details of this speculative model are not that important but I wanted to give the reader a sense of the style of much of theoretical physics today.

Warp Drive and Zero Point Energy

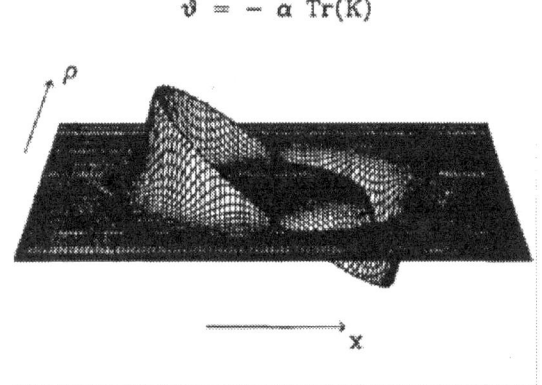

$$\vartheta = -\alpha \, Tr(K)$$

The Warp Drive: Hyper-Fast
Travel Within General Relativity
Miguel Alcubierre
Class. Quantum Grav. 11 (1994), L73-L77

The zero point energy density is positive with negative pressure on the left side of the above picture and it is negative with positive pressure on the right side. The ship moves to the right. There is a blue shift from the positive energy density in the stern and a red shift from the negative energy in the bow. This is opposite to the motional Doppler shift. We call this the "reverse Doppler shift' observed in UFOs by Dr. Bruce Cornet.

The warp drive: hyper-fast travel within general relativity
Authors: Miguel Alcubierre
Journal-ref: Class.Quant.Grav. 11 (1994) L73-L77

It is shown how, within the framework of general relativity and without the introduction of wormholes, it is possible to modify a space-time in a way that allows a spaceship to travel with an arbitrarily large speed. By a purely local expansion of space-time behind the spaceship and an opposite contraction in front of it, motion faster than the speed of light as seen by observers outside the disturbed region is possible. The resulting distortion is reminiscent of the "warp drive" of science fiction. However, just as it happens with wormholes, exotic matter will be needed in order to generate a distortion of space-time like the one discussed here.[lxxx]

Precision Cosmology's Equations

On the very large scale where our entire mighty galaxy is approximated as a mere speck, we assume the CP, no that's not Communist Party, it's Cosmological Principle that the universe is homogeneous and isotropic. The FLRW CP solution to Einstein's field equation

$$G_{\mu\nu} + \Lambda_{zpf} g_{\mu\nu} = -\frac{8\pi G}{c^4} T_{\mu\nu} \tag{1.6}$$

is in the, immensely practically useful choice of the Hubble flow coordinates

$$ds^2 = (cdt)^2 - R(t)^2 \left[\frac{dr^2}{1-kr^2} + r^2 \left(d\theta^2 + \sin^2\theta d\phi^2 \right) \right] \tag{1.7}$$

where $R(t)$ 3D space scale factor (with dimensions of length) for the now observed accelerating expansion rate of the universe from the antigravitating universally repelling "dark energy" seen in the redshifts of Type 1a supernovae "standard candles". Note that r is dimensionless, a pure number in the above convention. The name of this cosmology game is to get a simple partial differential equation for $R(t)$ by plugging the metric equation into the Einstein field equation. A smart high school kid with a natural aptitude for physics can do it. The inflation conjecture, that I now have a heuristic dynamics for as the vacuum condensation of bound state virtual electron-positron pairs, demands that 3D space is flat so that k = 0. This number is now measured to about 2% precision and that error is shrinking with new experiments.

The above-mentioned plugging in and using local conservation of mass-stress current densities gives, assuming that the matter tensor $T_{\mu\nu}$ is a perfect fluid, and assuming inflation

$$3\left(\frac{dR/dt}{R} \right)^2 \equiv 3H^2 = 8\pi G\rho + c^2\Lambda - \frac{3kc^2}{R^2} \tag{1.8}$$
$$k = 0$$

$$\frac{1}{R}\frac{d^2R}{dt^2} = -\frac{4\pi G}{3}\left(\rho + \frac{3p}{c^2} \right) + \frac{c^2\Lambda}{3} = -\frac{4\pi G}{3}\sum_i \rho_i\left(1 + 3w_i\right) \tag{1.9}$$

$$\frac{d\rho}{dt} = -3H\left(\rho + \frac{p}{c^2} \right) = -3H\rho(1+w) \tag{1.10}$$

$$w \equiv \frac{p}{c^2\rho} \tag{1.11}$$

World Stuff (both real and virtual)	w
Matter, i.e. real leptons & bound-states of several quarks	0
Radiation, i.e. real transverse polarized far-field photons	+1/3
Zero point vacuum fluctuations, i.e. random virtual quanta of all fields	-1
Near EM fields, i.e. macro-quantum coherent states of longitudinal polarized virtual photons	?

The phenomenological inflation model uses a simplified large-scale Landau-Ginzburg macro-quantum field equation for the ODLRO vacuum coherence that neglects local space variations in this CP limiting case.

$$\rho_{zpf} c^2 = \frac{c^4 \Lambda_{zpf}}{8\pi G} = \frac{\kappa}{2} \left| \frac{d\Psi}{dt} \right|^2 + V(\Psi, \Psi^*)$$

$$p_{zpf} = \frac{\kappa}{2} \left| \frac{d\Psi}{dt} \right|^2 - V(\Psi, \Psi^*)$$

(1.12)

Note that $d\Psi/dt \to 0 \Rightarrow w \to -1$.

The critical density of all stuff, both real and virtual exotic vacua at cosmic Hubble flow time t measured by the cosmic blackbody radiation temperature is

$$\rho_c \equiv \frac{3H^2}{8\pi G}$$

(1.13)

Define the dimensionless

$$\Omega_i \equiv \frac{\rho_i}{\rho_c}$$

(1.14)

Therefore, the first of the above Hubble flow equations is simply

$$\sum_i \Omega_i = 1$$

(1.15)

Creation of the curved Higgs Ocean from the flat Dirac Sea

In non-perturbative BCS transition, below the phase transition temperature, we know that you cannot get from the normal metal to the superconductor with a finite number of Feynman diagrams in perturbation theory any more than you can get the complete curved 1916 GR from a rigid flat Minkowski background. In the superconductor

Binding Energy Per Real Pair $\sim$ (Debye Energy)$e^{1/[(\text{Fermi Surface Density of Electron States Per Unit Energy})(\text{Interaction Energy})]}$ (Step Function)

Step Function = 0 when pair interaction energy > 0, i.e repulsive Coulomb force dominates, i.e. normal metal.

Step Function = 1 when pair interaction energy < 0, attractive virtual phonon force exceeds repulsive virtual photon force.

Note that as Interaction Energy $\to$ 0 from negative values, the binding energy also $\to$ 0 and obviously stays zero for repulsive interaction, i.e. $e^{-\frac{1}{0}} = e^{-\infty} \to 0$.

As the coupling $\to \infty$, the binding energy approaches the Debye energy. The Debye Energy is the width from the Fermi surface of the normal metal in which real electrons pair into a particular EPR entangled state with spin zero and opposite momenta in simplest case. OK it's the same general story for emergent gravity. "Fermi surface" is replaced by "Negative Energy Dirac Sea" with m = 0, i.e. Fermi energy = 0 in the false vacuum. The real electron pair of charge 2e is replaced by neutral virtual electron-positron pair. "Debye Energy" I thes Heisenberg energy uncertainty associated with QED vacuum polarization. The Debye energy in a crystal has to do with the strength of the electron-phonon coupling. Here it is from electron-photon coupling. The large-scale low-energy sector of the Higgs Ocean is from macroscopic occupation of the same single-particle center-of-mass "boson" wave function by jillions and jillions of virtual electron-positron pairs exchanging a virtual photon. The basic Feynman diagram is a Greek Theta "vacuum bubble".

Decoding The Cipher of Genesis

In The Beginning was The Question.

"The Question is: What is The Question?"

Einstein said: Let there be The Light Cone.

And there was The Light Cone.

Correct me if I am wrong, but so far there is no evidence for quantum gravity foam in high-energy cosmic rays and no evidence at all for exotic dark matter real particles. I predict, for the record, that these null results will stick as a matter of fundamental principle if Andrei Sakharov was right that gravity is emergent. What was missing from Sakahrov's 1967 idea was vacuum ODLRO coherence (Higgs Field with Goldstone Phase).

I say that the "Higgs Ocean"[lxxxi] field and the Einstein-Cartan tetrad field are not independent from each other as everyone else today assumes without question. I use Higgs field loosely and roughly mean also the inflation field created at 10^{-33} cm. The electro-weak Higgs field splits off at $\sim 10^{-16}$ cm to give rest mass to quarks and leptons. The tetrad field emerges bottom $\rightarrow$ up from the "stiff" world hologram long-range macro-quantum coherent "Goldstone Phase" of the Higgs Ocean field. This is a simplified single Goldstone phase toy model to illustrate the idea. With two Goldstone phases the curved tetrad 1-form need not be closed, hence it has a bona-fide non-vanishing 2-form that is essentially an "area flux density" leading to the Hawking-Bekenstein quantization of the area of the event horizon of a black hole and beyond that the 't Hooft & Susskind world hologram conjecture that all of 3D space is a holographic image projected from a DeSitter space-time "screen" that is $\sim 10^{122}$ bits in our Hubble sphere local Level I universe in the cosmic landscape.

When the robust phase of the Higgs Ocean field goes incoherent, then the non-trivial warp field part of the tetrad vanishes and we are back to the pre-inflationary unstable false vacuum of massless spin 1/2 & spin 1 quantum field theory of globally flat 1905 Special Relativity without any gravity and rest inertia. The symmetry group of this unstable vacuum is the 15-parameter conformal group of Roger Penrose's theory of massless twistors described in his book "The Road to Reality". Penrose may be correct that we do not need extra space dimensions. On the other hand Lenny Susskind argues that we do need them to explain the extraordinary fine-tuning of the cosmological constant to maybe the 120th decimal place. For this we need the 10^{500} vacua of the 6 extra space dimensions of the Calabi-Yau fiber. Such a complex vacuum landscape together with chaotic eternal inflation allows us to come into being in the sense of the Weak Anthropic Principle. We also know that warp drive and star gate time travel to the past need torsion fields to extend Einstein's 1915 curvature theory. We cannot conquer space and time by curvature alone. We need torsion too. Well torsion fields are another way of getting the 6 extra space dimensions! The nice feature of extra space dimensions, however, is that it can give us the Abdus-Salam strong short-range gravity $G* \sim 10^{40}G$ that we need to model the lepto-quarks as Wheeler "mass without mass" and "charge without charge" micro-geon Kerr-Newman-Burinski solutions of pure vacuum with exotic cores of dark matter on the scale of 1 fermi $10^{-13}cm$ or less. We may not need supersymmetry either.

That's the heuristic proof of no gravitons and no quantum foam![lxxxii]

"Maybe not as foamy as some scientists thought, as a fresh look at a quasar first observed in 1998 by the Hubble Space Telescope (HST) shows. Physicists observed a diffraction pattern called an Airy ring around the image of a distant quasar-like object. This ring persuades physicists that the light from this distant object has traveled through a relatively calm – rather than extremely frothy – spacetime. ... Ng and his colleagues, W.A. Christiansen and H. van Dam from the University of North Carolina, have narrowed down the possible models of quantum foam into the least foamy variations. The team compared two spacetime foam models based on quantum fluctuations in spacetime geometry. The first model is consistent with the holographic principle, which stipulates that the maximum amount of information that any region of space can store is proportional to its surface area instead of its volume, like a hologram. The second model, called the random-walk model, stipulates that successive fluctuations are random, totally uncorrelated. The holographic model enables a less turbulent spacetime compared with the random-walk model, which involves greater fluctuations. ... The HST's high-resolution image of the Airy ring surrounding the quasar ruled out the random-walk model, but lacked sufficient resolution to test the holographic model. ... 'This line of reasoning strongly hints at the existence of dark matter and dark energy, independent of the evidence from recent cosmological observations.' Although the universe may not be quite as foamy as some scientists previously suspected, the team has put constraints on the foaminess of spacetime, and supplied another parameter with which to probe unconventional matter and energy." By Lisa Zyga, *PhysOrg.com*

The inflationary critical phase transition Planck temperature for the Diff(4) covariant 4D Super-Solid Aether to form, like a superconductor from normal metal out of the unstable false flat Minkowski vacuum is

$$T_{critical} = \frac{hc}{k_B L_p}$$

$$L_p^2 \equiv \frac{Gh}{c^3}$$

The Bekenstein-Hawking world hologram "entropy" of the Universe, if you believe that model that George Chapline rejects, would be something like

IT FROM BIT

$$S_{Universe} = k_B R(t)^2$$

$$R\left(-13.7 \times 10^{13} \, years\right) \sim 0\left(L_p\right)$$

$$S_{Universe}\left(Beginning\right) = 1 \, BIT$$

This might explain the origin of the irreversible Arrow of Time, i.e. why we age as the 3D space of the universe expands and accelerates after a finite time of a few billion years from the hot Big Bang. On the other hand Gott has a different explanation based on the self-creating universe with time travel to the past. This problem is still not well understood.

In my model, emergent gravity totally disappears at the Planck phase transition! All inertia disappears. We are back to the pre-inflationary unstable chaotic globally flat Minkowski vacuum. All rest masses disappear! That's why quantizing g_{uv} is *unrenormalizable*. No gravitons and no quantum gravity foam is to be found. That idea was Wheeler's greatest blunder. Great men make great mistakes and lesser men waste their careers chasing after them.

All attempts to re-quantize emergent ODLRO fields fail, e.g. quantum gravity is unrenormalizable!

"The Question is: What is The Question?" Wheeler

Therefore, all attempts at "quantum gravity" must fail!

Why? It's the wrong question!

Matt Visser's Thoughts on Emergent Gravity

Visser is an expert on Star Gates. His book "Lorentzian Wormholes" was the first important book on the subject following the original paper on the subject by Kip Thorne and his students at Cal Tech in the mid 1980's.

Einstein gravity as an emergent phenomenon?[lxxxiii]

Carlos Barcelo and Matt Visser & Stefano Liberati

"In this essay we marshal evidence suggesting that Einstein gravity may be an emergent phenomenon, one that is not "fundamental" but rather is an almost automatic low-energy long-distance consequence of a wide class of theories. Specifically, the emergence of a curved spacetime "effective Lorentzian geometry" is a common generic result of linearizing a classical scalar field theory around some non-trivial background. This explains why so many different "analog models" of general relativity have recently been developed based on condensed matter physics; there is something more fundamental going on. Upon quantizing the linearized fluctuations around this background geometry, the one-loop efective action is guaranteed to contain a term proportional to the Einstein–Hilbert action of general relativity, suggesting that while classical physics is responsible for generating an "effective geometry", quantum physics can be argued to induce an "effective dynamics". This physical picture suggests that Einstein gravity is an emergent low-energy long-distance phenomenon that is insensitive to the details of the high-energy short-distance physics. ... There is a risk that all current attempts at "quantizing gravity" are condemned to failure because they have been starting from a fundamentally flawed premise, and that in reality there are no fundamental gravitational degrees of freedom to quantize — it is possible that Einstein gravity is an emergent phenomenon, in the same sense that fluid dynamics emerges from molecular physics as a low-momentum long-distance approximation. In this essay we will take a careful look at this idea, and highlight some of the possibilities, problems, and opportunities that such a situation entails. ... We were led to these notions via current research on analog models of general relativity. Because of the extreme difficulty (and inadvisability) of working with intense gravitational fields in a laboratory setting, interest has now turned to investigating the possibility of simulating aspects of general relativity — though it is not a priori expected that all features of Einstein gravity can successfully be carried over to the analog models. Numerous rather different physical systems have now been seen to be useful for developing analog models of general relativity. A literature search as of March 2001 finds well over a hundred scientific articles devoted to one or another aspect of analog gravity and effective metric techniques. The sheer number of different physical situations lending themselves to an "effective metric" description strongly suggests that there is something deep and fundamental going on.

Typically these are models of general relativity, in the sense that they provide an effective metric and so generate the basic kinematical background in which general relativity resides; in the absence of any dynamics for that effective metric we cannot really speak about these systems as models for general relativity. However, as we will discuss more fully bellow, quantum effects in these analog models might provide some sort of dynamics resembling general relativity."

10 dimensional space-time of the Physical Vacuum

6 rotational coordinates (local O(1.3) group)

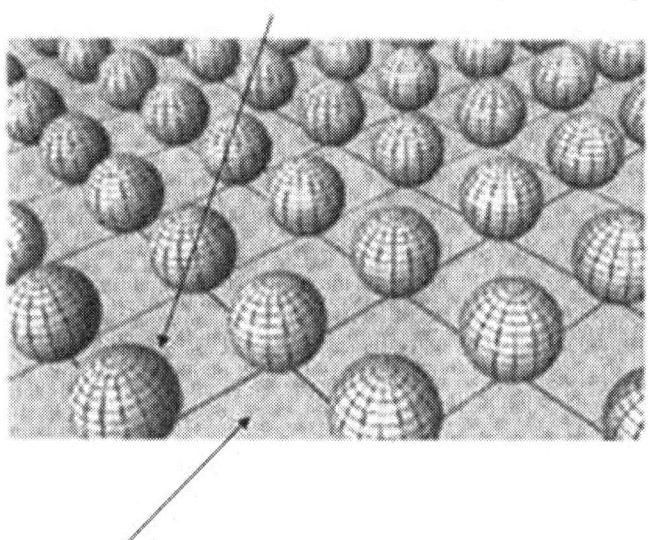

4 translational coordinates (local T(4) group)

Gennady Shipov's picture agrees with my theory.

In my theory the dynamics is from the local gauging of the translation space-time symmetry group T4 in same way that Maxwell's field theory is from the local gauging of the internal symmetry group U(1). I get quantization of area and holography from a simple model of the inflation vacuum condensate field with only two Goldstone phases from hidden internal symmetry in the post-inflation vacuum giving quantized 2D Bohr-Sommerfeld period integrals for a closed geometrodynamic area Cartan 2-form that is inexact when the interior has point defects in the inflation field as shown in my emergent gravity paper http://arxiv.org/abs/gr-qc/0602022 . If we had three Goldstone phases with four real scalar fields we could have a closed inexact volume 3-form, but this would contradict the area quantization and holography conjectures. Remember closed inexact Cartan forms have memory. Integrating them is path-dependent or anholonomic.

BCS Nonperturbative Dynamics from Microscopic Substratum's "False Vacuum"

As preparation for the off-mass-shell electrically neutral vacuum elastic super-solid of virtual electron-positron pairs, let us review the BCS model[lxxxiv] of the on-mass-shell electrically charged supercurrents of real electron pairs in a metal. Consider a pair of real electrons in a thin energy shell that interact above a "quiescent Fermi sphere" exchanging both virtual photons and virtual phonons from the crystal lattice such that the virtual phonon exchange is attractive and overpowers the repulsion between like charges from the virtual photon exchange. The non-perturbative BCS model gives the nonanalytic formula for the macroscopic eigenvalue N_0 of the electron-pair reduced quantum density matrix

$$N_0 \sim 2\rho\left(E_f\right)\hbar\omega_D e^{-1/\rho\left(E_f\right)U_{k_f}} \gg 1$$

$$U_{k_f} \sim -\frac{\left|M_{k_f}\right|^2}{\hbar\omega_D} + \frac{4\pi e^2}{V k_f^2} < 0 \qquad (2.1)$$

$$\left|\rho\left(E_f\right)U_{k_f}\right| \ll 1$$

The first term in the middle equation of (3.1) is the attractive electron-phonon Frohlich interaction. The positive term is the repulsive Coulomb interaction. The total number of pairs in the ground state condensate is N_0, $\rho\left(E_f\right) > 0$ is the density of electron states per unit energy at the Fermi surface of energy E_f and momentum k_f. The physical volume occupied by the electron pair is V. The effective spatial Fourier component of the interaction potential energy at the Fermi wave vector is U_{k_f}. The tiny energy gap difference $\Delta \sim k_B T_c$ per electron pair between the "false vacuum" normal metal ground state and the "true vacuum" superconducting ground state is the binding energy of the pair and critical temperature T_c that is given in

$$\frac{\Delta}{\hbar\omega_D} \sim e^{1/\rho\left(E_f\right)U_k} \ll 1$$

$$\rho\left(E_f\right)U_k < 0 \qquad (2.2)$$

The "normal fluid" random quasi-particle elementary excitations above the superconducting ground state that will cause electrical resistance and weaken the exclusion of magnetic flux in the Meissner effect have the "mass shell" spectrum

$$\varepsilon_k \sim \Delta + \frac{\left(\hbar\vec{k}\right)^2}{2\Delta/c^2} + \dots \qquad (2.3)$$

One can see heuristically how to make the analogy with the globally flat relativistic false vacuum of massless virtual negative energy electrons filling a Fermi sphere in a world crystal lattice of unit cell size $\sim L_p^*$ where the virtual electron-positron pair interaction is already attractive, to get

$$N_{e^+e^-} \sim 2\rho\left(E_f\right)\frac{hc}{L_p^*}e^{-1/\rho\left(E_f\right)U_{k_f}} \gg 1$$

$$U_{k_f} \sim -L_p^*\frac{\left|M_{k_f}\right|^2}{hc} - \frac{4\pi e^2}{L_p^{*3}k_f^2} < 0 \tag{2.4}$$

$$\left|\rho\left(E_f\right)U_{k_f}\right| \ll 1$$

$$E_f = 0$$

$$m_e \sim \frac{h}{cL_p^*}e^{1/\rho\left(E_f\right)U_k} \sim \frac{1}{2}10^6 ev \tag{2.5}$$

$$\varepsilon^2 = \left(m_e c^2\right)^2 + \left(h\overset{\Gamma}{k}\right)^2 \tag{2.6}$$

This is a primitive toy model Higgs mechanism where the "electron" gets its gravitational inertia (rest mass) from the vacuum coherence where

$$\Psi_{e^+e^-} \approx \sqrt{2\rho\left(E_f\right)\frac{hc}{L_p^*}e^{-1/\rho\left(E_f\right)U_{k_f}}}\ e^{i\arg\Psi_{e^+e^-}} \tag{2.7}$$

Nonlocality of the Gravity Energy

On Feb 4, 2006, at 5:15 AM, Waldyr A. Rodrigues Jr. wrote:

Dear Jack, … it is impossible to find any trustworthy energy-momentum conservation law in that theory. I recall what Sachs and Wu wrote in their book (page 97): *General Relativity for Mathematicians* (Springer, 1977): *'As mentioned in Section 3.8, conservation laws have a great predictive power. It is a shame to lose the special relativistic total energy conservation law (Section 3.10.2) in general relativity. Many of the attempts to resurrect it are quite interesting: many are simply garbage.'* … Details on the reasons why conservation laws did not exist in general in GR models and also in teleparallel version of GR (and more generally in Einstein-Cartan type theories) can be found in my new book: W. A. Rodrigues Jr. and E. Capelas de Oliveira, *The Many Faces of Maxwell, Dirac and Einstein Equations. A Clifford Bundle Approach.*[lxxxv] The formula for the metric field given in exponential form, given by Yilmaz (Eq.(2.6.)) in his paper *Field Theory of Gravitation,*[lxxxvi] involves the exponentiation of a symmetric tensor field and its trace. So, it is a very general metric field. However, Yilmaz did *not* solve the energy-momentum conservation law in his theory because it is based in a curved spacetime as is GR and so the same arguments that forbid the existence of energy-momentum conservation law in GR apply to Yilmaz theory."

Roger Penrose in his book "The Road to Reality" argues that the pure gravity vacuum energy is "nonlocal."

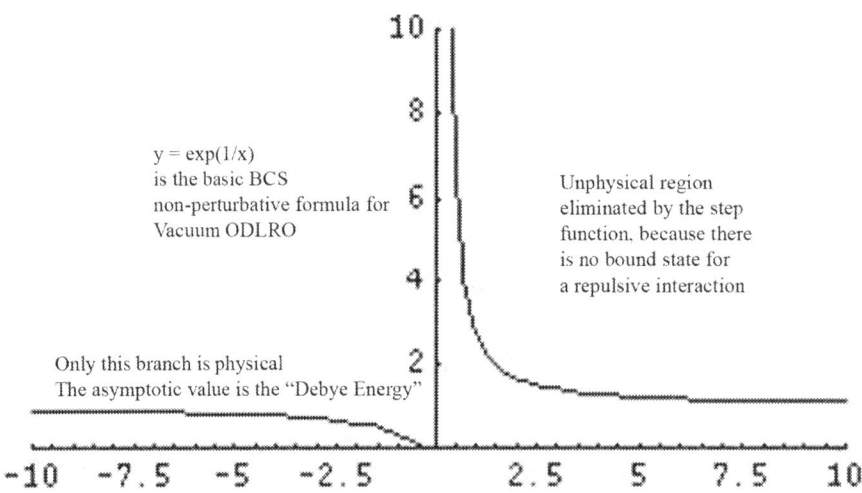

The hot Big Bang is powered by the binding energy per virtual pair bound state (normalized to "Debye energy" thickness of Fermi sphere in which most of the entangled EPR correlated virtual pairs originate) ~ quasi-particle rest mass of the lepto-quarks. That binding energy released in inflation, where jillions of virtual pairs Bose-Einstein condense into the same macroscopically-occupied single-particle center-of-mass quantum state, is y where

$$y = e^{\frac{1}{x}}\Theta(-x)$$

$\Theta(-x)$ is the stepfunction(-x) = SUM of infinity of Feynman vacuum polarization diagrams without external on-mass-shell lines of a certain class of virtual fermion-antifermion pairs exchanging virtual bosons in the false vacuum

It is the non-perturbative "BCS"(Nambu-Jona Lasino) basic formula that applies generally here.

$$y = |VacuumODLRO|$$

x ~ density of states per unit energy at Fermi surface of zero energy

$$d\rho_{virtuale^-}(E = 0)/dE$$

in the conformal massless gravity-free false vacuum multiplied by the interaction first order perturbation theory matrix element $\langle 0 | j_{e^-}{}^{\mu} D_{\mu\nu} j_{e^+}{}^{\nu} | 0 \rangle$ between the virtual fermion-antifermion pair that forms the Vacuum ODLRO condensate in the inflation phase transition.

$$\lim_{x \to 0^-} y \to 0$$

$$\lim_{x \to 0^+} y \to +\infty$$

Paul Zielinski on May 5, 2005: *"But this "energy problem" is a *foundational* issue, not a purely mathematical question."*

Indeed, I have outlined what I think may be the ultimate resolution of *why* gravity vacuum energy is nonlocal. The global DeRham integrals of a local "zero" density need not vanish in a multiply-connected region (non-trivial cohomology) of integration (peppered with little star gates) in some cases because of the "holes" that are singularities in the single-valued Vacuum ODLRO out of which the local GR field equations emerge. The nonlocality of the gravity energy is a strong "Cosmic Trigger" signal that Einstein's GR is an emergent c-number IR effective field theory from vacuum ODLRO! Andrei Sakharov did not have the right idea when he thought that GR emerges from the *random* ZPF. That is wrong, rather Einstein's gravity emerges from the *cohering* of the random ZPF in the *inflationary* phase transition of the false flat vacuum to the stable curved vacuum. This led Puthoff, Haisch & Rueda & Co off on a doomed wild goose chase for the "origin of inertia" using the wrong SED theory of Trevor Marshall. The nonlocality of gravity energy is an undecidable Godel question that requires a larger covering theory, i.e. Vacuum ODLRO. My supersolid paper of 1969 was the beginning of this idea as noted by George Chapline Jr.

The Initial Singularity

There is a lot of nonsense that "The Big Bang Never Happened". Of course it happened and is still happening with baby parallel universes in the multiverse of chaotic inflation. And that is why there is an Arrow of Time in which we age as space expands and indeed is now accelerating in its speed of expansion from cosmic antigravity that is zero point energy density of negative pressure. But where is the singularity in our own local Hubble sphere universe that Max Tegmark calls "Level I"? It is obviously the inflation vacuum phase transition $\rightarrow$ hot Big Bang (from the release of binding energy of virtual massless fermion-antifermion pairs). Lenny Susskind's "Cosmic Landscape" is equivalent to Max Tegmark's Levels I & II:

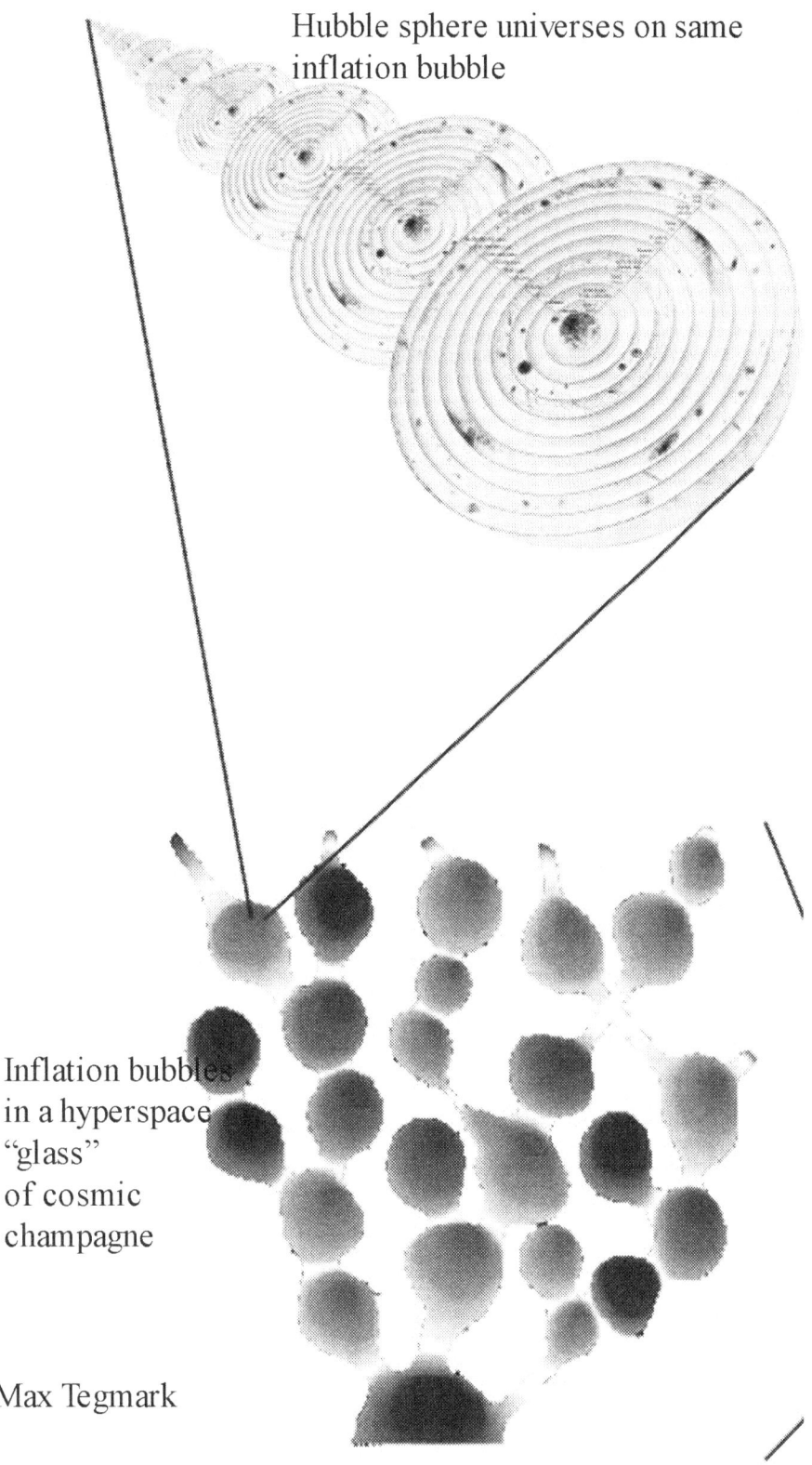

Hubble sphere universes on same
inflation bubble

Inflation bubbles
in a hyperspace
"glass"
of cosmic
champagne

Max Tegmark

*"Level I: A generic prediction of inflation is an infinite ergodic universe, which
contains Hubble volumes realizing all initial conditions - including an identical copy of*

you about $10^{\{10^{29}\}}$ meters away.

Level II: In chaotic inflation, other thermalized regions may have different effective physical constants, dimensionality and particle content." Max Tegmark[lxxxvii]

Spontaneous symmetry breaking

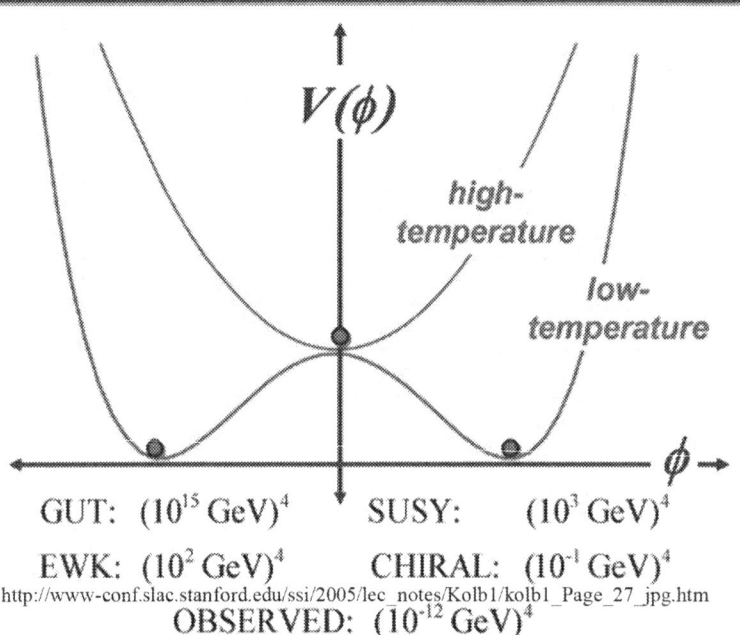

http://www-conf.slac.stanford.edu/ssi/2005/lec_notes/Kolb1/kolb1_Page_27_jpg.htm

"High temperature" in our particular application of the general idea of the emergence of new order should be replaced by "pre-inflation unstable false flat vacuum without gravity and inertia" and "low-temperature" by "post-inflation meta-stable curved vacuum with gravity and inertia." The standard model of leptons and quarks with electromagnetic, weak and strong local gauge forces, and its GUT to M-Theory speculative extensions, has several of these vacuum ODLRO macro-quantum fields as candidates to make the inflation field from in the initial Planck scale phase transition nucleating a single Level II inflation bubble that has an infinity of parallel Level I Hubble sphere universes. Hyperspace has an infinity of these Level II inflation bubbles. So we have an infinity, or, more precisely, an infinity of infinite sets of parallel universes. The chiral QCD coherent field provides most of the inertia of the quarks as kinetic energy trapped inside "bags". The smaller rest mass of the quarks comes from the $U(1)_{hypercharge}$ SU(2) vacuum symmetry breaking. How do we choose the vacuum condensate in my theory? I suggested the binding of massless virtual electron-positron pairs in the false vacuum. No one knows exactly how to do this yet. Of course, the condensate for emergent gravity must be neutral in all charges. That is the case for the component of the weak-hypercharge component that develops a macro-quantum coherent vacuum expectation value (VEV). The new insight here is that Einstein's 1915 field equations emerge simply from the coherent rigid phases of the cohering of the zero point vacuum fluctuations from all the quantum fields in the massless unstable globally flat pre-inflationary false vacuum.

Electron Stability & Rotating Dark Energy Charge Clusters

A dark matter core of negative zero point energy density with positive pressure, assuming w = - 1, so that $\Lambda_{zpf} < 0$ will stabilize a shell of unbalanced electric charge Ne explaining why the single electron is stable when $N = 1$. [lxxxviii] However, here for large N

$$V = V_{zpf} + V_{Coulomb} + V_{Rotation} = -c^2\Lambda_{zpf}r^2 + \frac{\left((Ne)^2 + \varsigma hc\right)}{Nmr} + \frac{J^2}{N^2m^2r^2} \quad (1.1)$$

$$(Ne)^2 ? \varsigma hc$$

The critical point for stability in the simple approximate spherically symmetric toy model with rotation is

$$\frac{\partial V}{\partial r} = 0$$

$$-2c^2\Lambda_{zpf}r - \frac{Ne^2}{mr^2} - 2\frac{J^2}{N^2m^2r^3} = 0 \quad (1.2)$$

The static limit stability condition is

$$\frac{\partial^2 V}{\partial r^2} > 0 \quad (1.3)$$

That in our model is

$$-2c^2\Lambda_{zpf} + 2\frac{Ne^2}{mr^3} + 6\frac{J^2}{N^2m^2r^4} > 0 \quad (1.4)$$

Furthermore, impose the close packing constraint

$$N\left(\frac{h}{mc}\right)^2 = 4\pi r^2$$

$$r = \sqrt{\frac{N}{4\pi}}\left(\frac{h}{mc}\right) \quad (1.5)$$

Solve for Λ_{zpf}

$$-2c^2\Lambda_{zpf}\sqrt{\frac{N}{4\pi}}\left(\frac{h}{mc}\right)-\frac{4\pi e^2}{m\left(\frac{h}{mc}\right)^2}-2\frac{h^2}{m^2\sqrt{\frac{N}{4\pi}}^3\left(\frac{h}{mc}\right)^3}=0$$

$$\left|\Lambda_{zpf}\right|=\frac{4\pi e^2}{2mc^2\sqrt{N}\left(\frac{h}{mc}\right)^3}+\frac{16\pi^2}{N^2\left(\frac{h}{mc}\right)^2}$$

$$=\frac{2\pi}{\sqrt{N}}\left(\frac{r_e}{\lambda_e^3}\right)+\frac{16\pi^2}{N^2\lambda_e^2} \qquad\qquad (1.6)$$

$$=\frac{2\pi}{\sqrt{N}}\left(\frac{\alpha}{\lambda_e^2}\right)+\frac{16\pi^2}{N^2\lambda_e^2}$$

$$=\left(\frac{2\pi\alpha}{\sqrt{N}}+\frac{16\pi^2}{N^2}\right)\left(\frac{1}{\lambda_e}\right)^2$$

$$\alpha\sim\frac{1}{137}$$

String Theory Without String Theory

Emergent Gravity

J. Sarfatti
ISEP Corporation
San Francisco, CA
(Revised 12th Draft, updates to this work in progress posted at
http://arxiv.org/abs/gr-qc/0602022)

Abstract

I derive Einstein's 1915 classical field theory of gravity with what resembles both a massive torsion field and the Calabi-Yau degrees of freedom from a conjectured eight Goldstone phases of the cosmic inflation field provided that the full Poincare group is locally gauged and its Lorentz subgroup is spontaneously broken in the vacuum. What looks like both the 't Hooft & Susskind world hologram conjecture of "volume-without-volume" and the quantization of area in Planck units given by Bekenstein and Hawking seem to be natural consequences of the conjecture. Just as the Michelson-Morley experiment gave a null result, this model predicts that the LHC will *never* find any viable dark matter exotic particles as a matter of fundamental principle -- neither will any other conceivable dark matter detector.

The intuitive physical motivation

The superfluid velocity field in Galilean relativity is the 3D Cartan closed inexact 1-form

$$v = \left(\frac{h}{m}\right) d\Theta \qquad (1.7)$$

Where the 0-form Θ is the generic "Goldstone phase" of the single-valued ODLRO ground state degenerate S^1 local order parameter manifold with quantized circulation. The quantization of area in geometrodynamics has a similar origin, but for a closed inexact 2-form. The stable topological defects where the order parameter vanishes for the closed inexact 1-form are line vortices and the closed loop line integral of the superfluid velocity field is quantized with an integer number of circulation quanta that correspond to vorticity flux quanta inside the vortex core whose diameter is a coherence length ξ -- similarly for magnetic flux quanta in Type II superconductors. Indeed there may be an analogous global topological effect in self-trapped laser filaments. Anyons in 2D thin films are electrons stuck on magnetic flux quanta with an effective fractional charge and a new kind of quantum statistics interpolating between fermions and bosons as shown by Frank Wilczek, Robert B. Laughlin et-al.[lxxxix] They are another example of interesting topological physics. Andrei Sakharov suggested that gravity was emergent from zero point vacuum fluctuations and some progress has been made along these lines by S. Adler and others. However, the arguments are little more than imposing a cutoff at the Planck scale. This paper takes the view that Sakharov asked the wrong question in 1967. We do not expect smooth classical gravity to emerge from the empirically small locally incoherent residual zero point fluctuations of the dark energy, but rather to emerge from a coherent vacuum field. The phenomenological inflation field is such a generic "Higgs" type of field. Although, it is taken as homogeneous in the FRW cosmological limit we extend it as a local field and show that Einstein's theory of gravity as curvature is naturally associated with the variations in its Goldstone phases and need not be assumed independently. As a first step we note that a model with two effective Goldstone phases seems to give area quantization and the volume without volume of the hologram conjecture. However, this turns out to be a degenerate case of a more natural theory with eight Goldstone phases in which the additional six are associated with the torsion field model suggested by Gennady Shipov in Moscow and Bangkok as well as the Calabi-Yau extra space dimensions provided that the Lorentz group is both locally gauged in the action and spontaneously broken or "hidden"[xc] in the physical vacuum giving massive torsion quanta. These Meissner effect masses $M_i, i = 1, 2, ...6$ determine the compactification scales

$$R_i \approx \frac{h}{M_i c} \qquad (1.8)$$

The six extra space dimensions of Calabi-Yau space are required in order to consistently derive the four curved tetrad fields from the eight independent Goldstone phases of the Planck-scale inflation field. This seems to give a natural reason, i.e., the local gauging and hiding of the Lorentz group, for why string theory requires the extra six space dimensions. Furthermore, these six space dimensions seem to be a projection from a seven dimensional space as suggested by the 11-Dim M-Theory conjecture. The equivalence principle in its deepest form seems to be the basic organizing idea required for M-Theory. By this I mean that all universal continuous space-time symmetries must be locally gauged for all physical dynamical actions. Einstein's 1915 General Theory of Relativity is simply the local gauge theory from the four-parameter translation space-time symmetry group of Special Relativity in this point of view first set forth by T.W.B. Kibble.[xci]

Short review of Einstein-Cartan tetrad gravity theory

The *four* Einstein-Cartan tetrad 1-form gravitational fields are

$$\varepsilon^a = I^a + |^a$$

$$\varepsilon^a = \varepsilon^a_\mu dx^\mu \tag{1.9}$$

$$a=0,1,2,3$$

The tetrad components ε^a_μ are local c-number fields in the "local coincidence" Diff(4) orbits P that are first-rank tensors in the Greek indices under the 4-parameter Diff(4) and are also first-rank tensors in the Latin indices under the 6-parameter Lorentz group. It is always possible to find transformations in which a geodesic LIF is momentarily aligned with a locally coincident non-geodesic LNIF in which

$$I^a_\mu \to \delta^a_\mu$$

$$I^a = I^a_\mu dx^\mu \tag{1.10}$$

Einstein's local equivalence principle is formally

$$g_{\mu\nu}(P) = \varepsilon^a_\mu(P)\eta_{ab}\varepsilon^b_\nu(P)$$

$$= \left(I^a_\mu(P) + |^a_\mu(P)\right)\eta_{ab}\left(I^b_\nu(P) + |^b_\nu(P)\right)$$

$$= I^a_\mu \eta_{ab} I^b_\nu + I^a_\mu \eta_{ab} |^b_\nu + |^a_\mu \eta_{ab} I^b_\nu + |^a_\mu \eta_{ab} |^b_\nu \tag{1.11}$$

$$\eta_{ab} = \begin{matrix} 1 & 0 & 0 & 0 \\ 0 & -1 & 0 & 0 \\ 0 & 0 & -1 & 0 \\ 0 & 0 & 0 & -1 \end{matrix}$$

Note the nonlinear "plastic" fracture[xcii] "geon" term quadratic in the curved tetrad field $|$. When $| = 0$ over a space-time region, applying Diff(4) still gives a *pseudo-curved metric* from the LNIFs created by the applied electromagnetic forces on detectors placed on non-geodesic paths. The litmus test of tidal geodesic deviation using pairs of test particles will always give a null result in such a case

The exterior Diff(4) covariant derivative is defined as

$$D = d + \Sigma^a_b \wedge \tag{1.12}$$

Where the six torsion-free spin-connection 1-forms are

$$\Sigma^{ab} = \Sigma^{ab}_\mu dx^\mu$$

$$\Sigma^{ab} = -\Sigma^{ba} \tag{1.13}$$

$$\Sigma^{ab} = \Sigma^{ab}_\mu dx^\mu$$

Einstein's 1915 theory has zero torsion, therefore the six spin connection 1-forms Σ^{ab} are constrained by the four curved tetrad 1-form fields I^a. The spin connection in this zero torsion field limit has no independent dynamical torsion field degrees of freedom. The four 1-form curved tetrad fields I^a come from locally gauging the global translation subgroup T_4 of the Poincare group. Dynamically independent torsion fields would require an additional local gauging of the local Lorentz group $SO(1,3)$. The six tidal geodesic deviation curvature 2-forms are

$$R^{ab} = D\Sigma^{ab} = d\Sigma^{ab} + \Sigma^a_\gamma \wedge \Sigma^{\gamma b}$$

$$R^{ab} = R^{ab}_{\mu\nu} dx^\mu \wedge dx^\nu \tag{1.14}$$

Many modern text books show how get to Einstein's 1915 theory with the Einstein-Hilbert action and zero torsion and we have nothing new to say about that, nor do we have anything noteworthy to say about the nonlocality of the pure gravity vacuum energy.

The new idea

All of the above is standard physics for introductory background. We now come to the original contribution of this paper. We first make a zero-order phenomenological toy model in which the Planck-scale inflation vacuum coherent field has three real scalar vacuum ODLRO *inflation field* components φ_i and obeys the usual renormalizable quartic Landau-Ginzburg generalized "Mexican Hat" c-number effective potential with spontaneous symmetry breaking prior to the kinds of quantum corrections discussed by Sidney Coleman at Erice, Italy.

$$V(\varphi) = \omega|\varphi|^2 + \kappa|\varphi|^4$$
$$\omega < 0, \kappa > 0 \tag{1.15}$$
$$|\varphi|^2 \equiv \varphi_1^2 + \varphi_2^2 + \varphi_3^2$$

The relative minimum in the landscape defining the G/H vacuum manifold is

$$\frac{\delta V}{\delta|\varphi|} = 2\omega|\varphi| + 4\kappa|\varphi|^3 = 0$$
$$\omega + 2\kappa|\varphi|^2 = 0 \tag{1.16}$$
$$|\varphi| = \sqrt{\frac{-\omega}{\kappa}}$$

Meta-stability of the post-inflationary vacuum requires that

$$\frac{\delta^2 V}{\delta|\varphi|^2} = 2\omega + 12\kappa|\varphi|^2 > 0 \tag{1.17}$$

Therefore

$$2\omega - 12\kappa\frac{\omega}{\kappa} > 0 \rightarrow -10\omega > 0 \tag{1.18}$$

Therefore, the degenerate vacuum manifold has the shape of the unit spherical surface in 2D with the *two independent* 0-form Goldstone phases as any two angles for the direction cosines. That is,

$$1 = \frac{\varphi_1^2 + \varphi_2^2 + \varphi_3^2}{|\varphi^2|} = \cos^2\theta + \cos^2\phi + \cos^2\chi \tag{1.19}$$

In general the 8 independent Goldstone phases are from the larger degenerate vacuum manifold

$$\frac{\sum_{i=1}^{9}(\varphi_i)^2}{|\varphi|^2} \equiv 1 = \sum_{i*=1}^{9}\cos^2\chi_i \qquad (1.20)$$

The additional 6 Calabi-Yau Goldstone phases are from spontaneously breaking the Lorentz group in the vacuum. Presumably, the initial two Goldstone phases are from spontaneously breaking some internal symmetry group in the Planck-scale inflation quantum phase transition. However, this is an open empirical question.

In addition, in order to connect torsion with Calabi-Yau, the full 10-parameter Poincare group is locally gauged to give 10 compensating gauge potential 1-forms of curvature and torsion.

$$A^0 \equiv |^0, A^1 \equiv |^1, A^2 \equiv |^2, A^3 \equiv |^3$$
$$A^4 \equiv \delta\Sigma^{13}, A^5 \equiv \delta\Sigma^{12}, A^6 \equiv \delta\Sigma^{23} \qquad (1.21)$$
$$A^7 \equiv \delta\Sigma^{01}, A^8 \equiv \delta\Sigma^{02}, A^9 \equiv \delta\Sigma^{03}$$

The curvature-torsion gauge potentials obey the Lie algebra of the Poincare group with structure constants $C_\gamma^{\alpha\beta}$:

$$\left[A^\alpha, A^\beta\right] = C_\gamma^{\alpha\beta}A^\gamma$$
$$A^\alpha = A_\mu^\alpha dx^\mu \qquad (1.22)$$
$$\alpha,\beta,\gamma = 0,1...9$$

The gauge covariant derivative on the ODLRO source field is

$$\varphi_{,\mu i} \equiv \partial_\mu\varphi_i + T_\alpha A_\mu^\alpha \varphi_i \qquad (1.23)$$

Where T_a are the 10 "charges" of the Poincare group Lie algebra in accord with my conjectured "extended equivalence principle" that all the generators of space-time symmetries induce geometrodynamic fields. From this point of view the basic tetrad theory is like the Yang-Mills spin 1 theory. However, the actual geometrodynamic objects are bilinear in the tetrads, which is why gravity is considered a spin 2 theory. The conjectured spin 2 "gravitons" would therefore be entangled pair states of the spin 1 quanta of the tetrad field. Yang-Mills spin 1 fields are renormalizable in contrast to spin 2 fields. Furthermore, Roger Penrose has shown how general relativity can be recast as a spinor theory.[xciii] The 2-component spinors are qubits. In terms of Wheeler's "IT FROM BIT," the geometrodynamic IT tetrad Cartan 1-form fields are 2 quBIT strings if you want to think of the *conjectured* megaverse of $\sim 10^{500}$ pocket universes as a Vast Active Living Intelligence System (P.K. Dick) non-classical computer. Indeed, the tetrad gravitational fields can be thought of as emerging from a multi-component Penrose virtual spinor pair vacuum condensate that is the inflation field in a way reminiscent of the BCS pairing in a conventional superconductor.

A topological defect is defined as a region of space-time where the local macro-quantum vacuum condensate order parameter $|\varphi| \to 0$ so that the set of complementary Goldstone phases is undefined at the defect. Nevertheless, the order parameter must be single-valued. This implies a generalized Bohr-Sommerfeld flux quantization rule closely related to the generalized Bohm-Aharonov effect for closed inexact Cartan forms integrated over closed cycles without boundaries that are not themselves boundaries due to non-trivial topology of the associated physical manifolds. The rule is that stable topological defects must have non-trivial homotopy groups larger than the identity group.[xciv] For example, a single real scalar field in 3 + 1 space-time has the vacuum manifold $S^0 \equiv \{\pm1\}$ with stable wall-domain defects. Two real scalar fields have the degenerate vacuum manifold with only one Goldstone phase Θ when the unstable incoherent vacuum with internal symmetry group G spontaneously breaks down to the normal subgroup H with the coset quotient space partition of non-overlapping equivalence classes

$$\frac{G(incoherent)}{H(coherent)} = S^1 \tag{1.24}$$

The dynamical action is still invariant under G so that the symmetry is not dynamically broken in the Lagrangian as in the Zeeman and Stark effects. More precisely, as emphasized by Sidney Coleman in his famous 1973 Erice Lecture this "More is different" (P.W. Anderson's phrase) emergence comes from a "hidden" or "secret" symmetry. The first homotopy group in this particular example is non-trivial equal to the integers

$$\Pi_1\left(S^1\right) = Z \equiv \{0, \pm1, \pm2, \dots \pm\infty\} \tag{1.25}$$

The corresponding stable topological defect is a string vortex line with a core scale "coherence length" ξ. The vortex line is surrounded by closed loops that have no boundaries but are themselves not boundaries because of the single Goldstone phase θ is undefined or "singular" there. The point of this classic example, of course, is that the 1-form electromagnetic vector potential A is closed but nonexact $B = dA = 0$ locally, i.e., there is no electric field and no magnetic field from the steady shielded electric current spiraling around the solenoid directly acting on the electrons in the interferometer. Therefore, there is no classical force acting on the electrons that can explain the shift in the fringes. This is an example of the direct physical reality of David Bohm's "quantum potential" formed from the sum of the two alternative Feynman histories for the charges in motion whether single electrons in a beam or supercurrents. Nevertheless, the set of possible loop integrals obeys

$$\left\{ \int_{Loop} A_{\mu} dx^{\mu} \right\} = Z \qquad (1.26)$$

A well-known example of this is the closed inexact Bohm-Aharonov 1-form where the single Goldstone phase Θ is expressed in terms of its two real scalar vacuum ODLRO fields φ_i.

$$d\Theta = \frac{\varphi_1 d\varphi_2 - \varphi_2 d\varphi_1}{\varphi_1^{\,2} + \varphi_2^{\,2}}$$

$$\Theta \equiv \tan^{-1} \frac{\varphi_2}{\varphi_1} \qquad (1.27)$$

where the integer $N \subset Z$ is the chiral winding number, i.e. the number of times the single Goldstone phase Θ winds around the ground state manifold $G/H = S^1$ for a single 2π circuit of a closed loop in physical 3D space that surrounds the vortex line defect in the case of the macro-quantum superconductor version of the orginal micro-quantum Bohm-Aharonov effect done with ordinary electron beams rather than supercurrents. Similarly in superfluid Helium 4 the closed non-exact 1-form is the superfluid velocity 3-vector and the corresponding quantum of flux is that of circulation equals vorticity multiplied by the enclosed area of the loop $\sim h/m$ in Galilean relativity. The third example of single Goldstone phase phenomena is the lattice of magnetic vortices in a Type II superconductor. Still a fourth example is the self-trapped laser filaments in nonlinear optics studied by the classic experiments of Ray Chiao at Berkeley.

In the above case of 3 scalar fields it is the second homotopy group that is non-trivial giving stable point "monopole" defects in the order parameter field.

$$\Pi_2 \left(S^2 \right) = Z \qquad (1.28)$$

The integral of suitably defined closed 2-forms around closed surfaces surrounding point defects are now non-exact, i.e.

$$\left\{ \iint_{Closed} A_{\mu\nu} dx^{\mu} \wedge dx^{\nu} \right\} = Z \qquad (1.29)$$

When the integer spherical wrapping number $N \subset Z$ (2D winding number) is ± 1 we have a hedgehog that seems to fit the NASA Pioneer Anomaly. If the 2-form is a natural geometrodynamic area flux density then we have a more natural picture of why areas should be quantized if the gravity field emerges from a single-valued vacuum condensate. Furthermore, if that area flux quantum corresponds to a closed 2-form then one sees that the natural local measure of volume is zero when viewed as the "divergence" exterior derivative of that 2-form. This intuitively suggests that the geometrodynamic volume is a holographic "image" projected from the quasi-boundary area of that volume. A well-known closed inexact "monopole" point defect 2-form is[xcv]

$$A \equiv L_p^2 \sin|\Phi| \, d|\Theta| \wedge d|\Phi| = \frac{\varphi_1 d\varphi_2 \wedge d\varphi_3 + \varphi_2 d\varphi_3 \wedge d\varphi_1 + \varphi_3 d\varphi_1 \wedge d\varphi_2}{\left(\varphi_1^{\ 2} + \varphi_2^{\ 2} + \varphi_3^{\ 2}\right)^{3/2}} \quad (1.30)$$

Note that the maximum number of real scalar fields that can fit into a space and still give simple stable topological defects with non-trivial homotopy seems to be the dimension of the space itself.

We now modify the above toy model with only *two* independent Goldstone phases to one with *eight* Goldstone phases. Define the four "diagonal" *non-closed* 1-form curved space-time tetrad fields and the six dynamical "off-diagonal" torsion 1-form potentials as

$$|^a = \sqrt{\frac{hG}{c^3}} \left[d\Theta^a \Phi^a - \Theta^a d\Phi^a \right]$$

$$| = |^a \otimes \partial_a \qquad\qquad (1.31)$$

$$\delta\Sigma^{ab} \sim d\Theta^a \Phi^b - \Theta^b d\Phi^a - d\Theta^b \Phi^a + \Theta^a d\Phi^b = -\delta\Sigma^{ba}$$

$$\delta\Sigma = \delta\Sigma^{ab} \otimes \partial_a \partial_b$$

Note that in this model curvature is not possible when either $h \to 0$ or $c \to \infty$ or both even when $G \neq 0$. The Higgs type model above with the two independent Goldstone phases corresponds to the two contracted 0-forms

$$\Theta \equiv \Theta^a \otimes \partial_a$$

$$\Phi \equiv \Phi^a \otimes \partial_a$$

$$|\Theta| = \sqrt{\Theta^a \Theta_a} \qquad\qquad (1.32)$$

$$|\Phi| = \sqrt{\Phi^a \Phi_a}$$

The Geometrodynamic Area Flux Density 2-form $A \equiv L_p^2 \sin|\Phi| \, d|\Theta| \wedge d|\Phi|$ is not exact globally when integrated over a closed 2D surface that surrounds a point "monopole" stable topological defect and the closed surface integral is quantized. Note that $\Omega \equiv dA = 0$ locally is the ghostly *volume-without-volume* geometrodynamic flux density that locally vanishes, but is nonlocally finite defined as the surface area integral with area c-bits equal to the Planck area. The generalized Bohm-Aharonov effect and the single-valuedness of the Higgs field vacuum order parameter gives a "volume-without-volume" 't Hooft & Susskind "world hologram" of quantized area flux. Therefore, when a non-bounding closed surface S without boundary surrounds at least one singular point defect where $|\varphi| \to 0$ and the two Goldstone phases are undefined, then I get the Hawking-Bekenstein quantized area

$$\iint_S A = \iint_S L_p^2 \sin|\Theta|\, d|\Theta| \wedge d|\Phi| = 4\pi L_p^2 N$$
$$N = 0, 1, 2, \ldots$$

(1.33)

Where N is the number of times the S^2 vacuum manifold G/H is "wrapped around" for a single "wrap around" of the corresponding S^2 closed surface surrounding at least one point neutral monopole defect in ordinary space. This is the 2D analog of the more familiar S^1 vortex line defects of superfluids and superconductors. When we have the six additional phases unlocked then we have all the elaborate structure of the Calabi-Yau spaces discussed by Susskind.

The cosmological constant and the cosmic landscape

Virtual quanta inside the vacuum do not have any direct quantum electrodynamic effects. They only have indirect effects like the Lamb shift. In contrast, the equivalence principle demands that any residual virtual quanta have direct gravitational effects. I suggest the following "two-fluid" relationship between the cosmological constant locally random zero point energy and the large-scale locally non-random vacuum condensate inflation field order parameter in the FRW metric limit

$$\Lambda \sim \frac{1}{L_p^2}\left[1 - |\varphi|^2\right] \tag{1.34}$$

Where the inflation field intensity $|\varphi|^2$ is normalized to a pure number. The vacuum energy is locally incoherent, but is nonlocally coherent in the sense of Einstein-Podolsky-Rosen entanglement. This explains why, for example, the phenomenological superfluid helium density is ~ 100% of the total density near absolute zero even though the condensate density is only ~ 10% of the total density. The smooth geometrodynamic structure, what Einstein called the "marble" is in the Goldstone phases of the vacuum ODLRO condensate. The residual zero point fluctuations from all fields are what Einstein called the "wood." We really have a "three-fluid" model because the on-mass-shell elementary excitations are the "normal fluid." Robert Becker[xcvi] independently introduced the idea of the nonlocal coherence of the locally incoherent vacuum energy. This would of course imply the usual enormous zero point energy density from quantum field theory inside the tiny cores of the stable topological defects. Although the zero point energy density is large the core sizes are small. This formula is consistent with Leonard Susskind's model of the "cosmic landscape" in which the height of the landscape is Λ that can even go negative. There are approximately 10^{500} vacua minima from the Calabi-Yau extra-dimensional fiber spaces. Susskind then explains the extreme fine-tuning of our pocket universe's $w \sim -1$ $\Lambda \sim 0.73\rho_c$ in terms of the idea of eternal chaotic inflation of "bubbles" with an infinity of Max Tegmark's "Level I" pocket universes on each bubble in an infinity of "Level II" bubbles in hyperspace. Presumably, the "coincidence problem" is "explained in the same way? Thus, the cosmic landscape of possible universes is fully occupied by actual pocket universes in Susskind's scheme that then applies the natural selection of the Weak Anthropic Principle. Susskind emphasized[xcvii] that Steven Weinberg showed in 1987 that the actually observed value of our Λ is needed to allow carbon life within narrow limits. Weinberg made this prediction 13 years before the Type 1a supernovae data began to be measured and understood.

Gerry Gilmore reported recent dark matter measurements at the UCLA Dark Matter 2006 Conference that I attended.[xcviii] The measured mass scale is ~ fifty million solar masses over a length scale of ~ three hundred parsecs. Assuming dark matter is made out of particles whizzing through space the Newtonian virial theorem seems to be the basis for Gilmore's quote of a "dispersion" of 9 km/sec. However, my model of short-scale dark matter as inhomogeneous negative zero point energy density of positive pressure with $w < -1/3$ replaces that 9 km/sec with a negative vacuum energy density that is approximately one million times larger[xcix] (in absolute value) than the large-scale positive vacuum "dark energy" density with $w \approx -1$ negative pressure. Joel Primack[c] mentioned that $w \approx 0$ Cold Dark Matter (CDM) is working well in understanding the current indirect astronomical observations. My model here mimics CDM for distant observers and seems to be an equally valid interpretation that will become more convincing if local detection of real on-mass-shell dark matter particles is never achieved.

In this theory we have conjectured that the dark matter is actually negative vacuum energy density of positive pressure with $w < -1/3$ that mimics $w = 0$ CDM for distant observers.[ci] The "sound speed" would obviously be zero if that were the case. The only difference between the negative pressure "dark energy" and the positive pressure "dark matter" phases of exotic vacuum would be the vacuum coherence factor $|\varphi|^2$ on different scales. Indeed, the prediction of the unified vacuum energy interpretation of both dark energy and dark matter is that neither the Large Hadron Collider (LHC), nor any other conceivable dark matter detector, will ever find any viable on-mass-shell dark matter particles as a matter of fundamental principle. That is, a null result as a matter of principle like the null result of the Michelson-Morley experiment that failed to detect the motion of the Earth through the Galilean aether.

Acknowledgments

I would like to thank Professor Waldyr Rodrigues, Jr of UNICAMP, Brazil and Robert Kiehn for pointing out mathematical ambiguities in the early versions of this paper that is really a progress report. The reader should not infer either Professor Rodrigues or Professor Kiehn endorse all of the ideas of this paper or that they think the mathematics is up to their rigorous standards. Indeed they do not. I remind the reader that important new physics has often been created with "bad mathematics" that is usually patched up later. A case in point is the history of the Feynman diagrams and the path integrals, which are still non-rigorous from the point of view of the top mathematicians. Indeed Feynman wrote that he did not know what unitarity was when Dirac first brought it up to him. Heisenberg did not know rigorous matrix math when he did his early work and Einstein had notorious problems with the tensor mathematics in his ten-year struggle with General Relativity. Hagen Kleinert's "world crystal" ideas were an early motivation for these new ideas. George Chapline Jr. and Pawel Mazur gave talks on vacuum ODLRO emergence of gravity at the Pacific Coast Gravity Meeting 2006 at Kavli ITP, UCSB where I also presented a short version of this paper. The approach of Chapline and Mazur (reported in March 11, 2006 New Scientist) is not the same as mine although there are some basic similarities there are also significant differences of detail. Chapline also says there will be no locally observable dark matter particles and his reasoning is, I think, consistent with my theory here as well. Chapline with Laughlin have a whole new theory of quantum criticality at the event horizon surface of a black hole that is not part of my theory here as far as I currently understand its implications. One major philosophical difference I have with both Chapline and Susskind, both of whom I knew personally back in the 1960's, deals with the Hawking-Susskind debate over information loss down a black hole. I was present at GR 17 in Dublin where Hawking threw in the towel to Susskind – perhaps prematurely. Both Chapline and Susskind agree with each other that micro-quantum theory with unitarity and the no-cloning theorem apply to the information flow through the black hole surface. I have my doubts because the Landau-Ginzburg ODLRO macro-quantum theory replaces the linear nonlocal unitary entangled micro-quantum Schrodinger equation with an emergent effective nonlinear local nonunitary macro-quantum Landau-Ginzburg type equation for the evolution of the emergent condensate. Of course the elementary excitations in and out of the condensate reservoir still obey the unitary orthodox micro-quantum rules subject to Born's probability interpretation and Feynman's rules when to add path amplitudes coherently and when not to. However, the von-Neumann projection postulate does not apply directly to the condensate because of what P.W. Anderson calls "generalized phase rigidity" of the local macro-quantum order parameter that in geometrodynamics is the space-time stiffness "string tension" $\sim c^4/G$.

Where we are now?

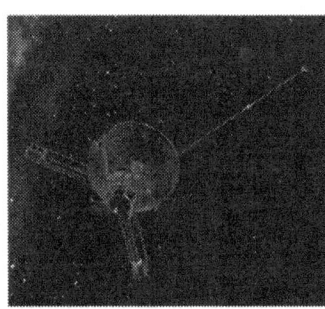

* Pioneer 10 & 11 anomaly in solar system (~70 AU):
  extra acceleration ~ 8.5×10^{-10} m s^{-2}

* Cosmological observations:
  72% of "dark energy"
  24% of "dark matter"
  4% of baryonic matter

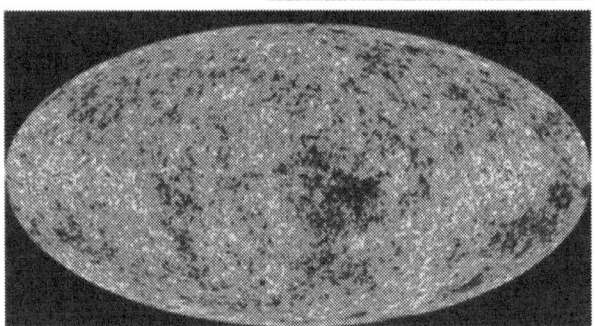

"photograph" of Universe at 380 000 years (now 13.7 billion years)

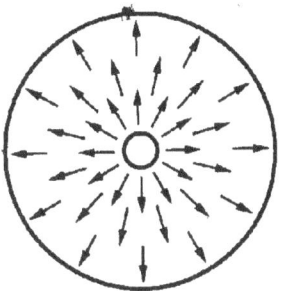

Fig. 1. Magnetization pointing outwards in the space between two spherical enclosing surfaces. This is known as a hedgehog. David Thouless Topological QM book

Explanation of NASA Pioneer Anomaly?

If in a hedgehog topological defect for the giant vacuum coherent field ϕ, the dark zero point energy density field obeys

$$\Lambda_{zpf}(r,t) = -\frac{H(t)}{cr}$$

$$H(t) = \frac{1}{a(t)}\frac{da(t)}{dt}$$

(1.35)

between two concentric spherical boundary surfaces centered on the Sun, the first surface at $\sim$ 20AU at approximately Jupiter's orbit, then

$$V_{zpf} = -c^2 \Lambda_{zpf}(r,t)r^2$$

$$a_{Pioneer} = -\frac{\partial V_{zpf}}{\partial r} = -cH(t) \qquad (1.36)$$

The Higgs vacuum manifold of ϕ out of which Einstein's gravity emerges must have the topology of SU(2) with two Goldstone phases defining the spherical surface vacuum manifold with a second homotopy group equal to the integers for this model to work.

What about the dark matter galactic halos seen in the flat stellar rotation curves?

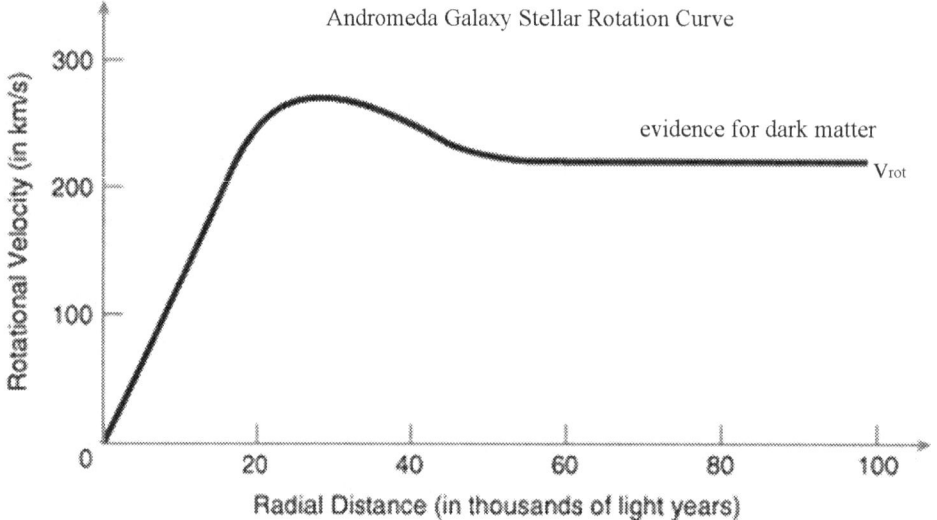

Andromeda Galaxy Stellar Rotation Curve

Galactic Halo?
If the dark matter is negative zero point energy density with positive pressure then we already gave a simple toy model above in which the zero point energy gradient provides the centripetal acceleration needed to cancel the centrifugal acceleration of the stars with roughly constant tangential rotational speed.

Dark matter detectors will never click with the right stuff to explain $\Omega_{DM} \sim 0.23$ if my explanation here is correct. Contrary to Lenny Susskind's expectation in "Cosmic Landscape" I propose that dark matter is not made from real particles on-mass-shell whizzing around space, but is simply the positive pressure phase of exotic vacuum.

WMAP: The World is Flat!

Note however that Lenny Susskind thinks that more precise future measurements will show some negative spatial curvature rather than flat, i.e. zero spatial curvature in the large-scale limit.

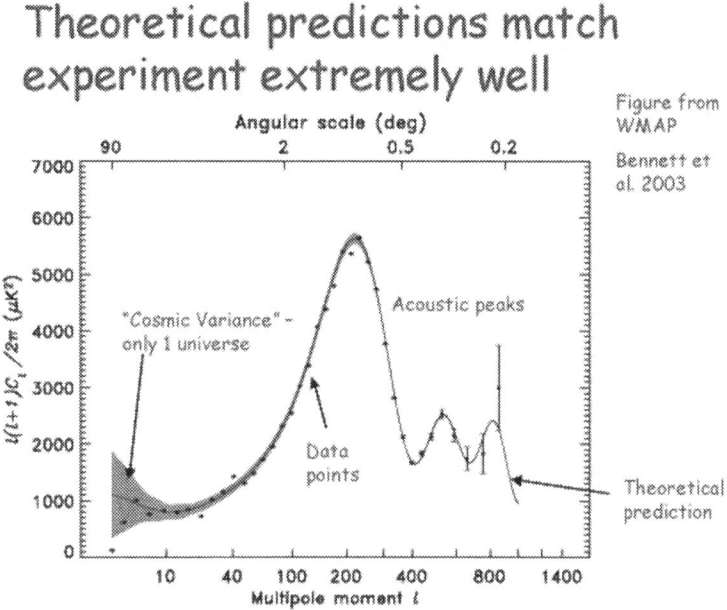

Tom Friedman has written a book on how the economic playing field has been leveled by the Internet weakening the American and European intellectual workers with competition from India and China. This economic deflation is in contrast to physical inflation that flattens 3D space on the very large scale, though not on the small scale. The WMAP data in the above slide is consistent with inflation where the space curvature vanishes, i.e. k = 0 in the standard homogeneous isotropic metric for very large scales used below.

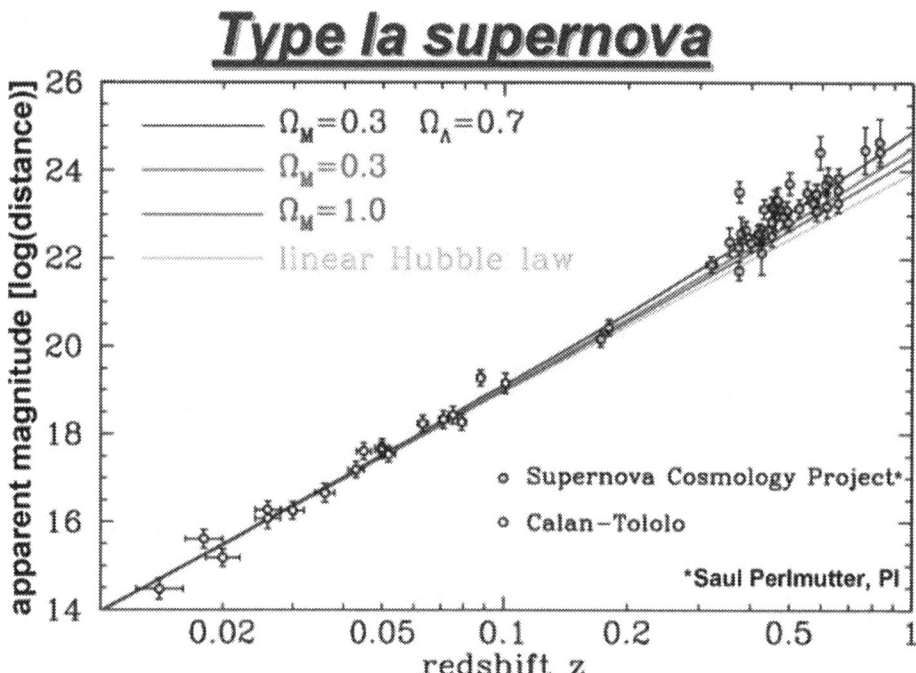

Two independent groups show that the standard candle Type 1a supernovae are dimmer than they should be at a given redshift z. This means that the speed of expansion of space is increasing. That is, we live in an accelerating universe from the repulsive anti-gravity influence of the negative quantum pressure of the positive zero point dark energy density uniformly and isotropically distributed all over space at large scales. The light from the Type 1a supernovae has spread out over a wider area than would be the case if the repulsive dark energy density was not out there. Therefore, the supernovae signals are weaker than was expected in the old model.

The numerical values of the key parameters observed in the past seven years or so are collected below from SLAC SSI 2005.

Precision cosmology

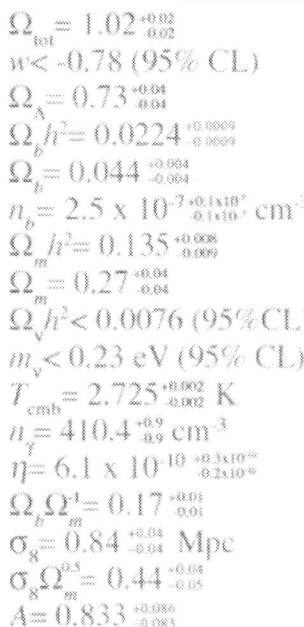

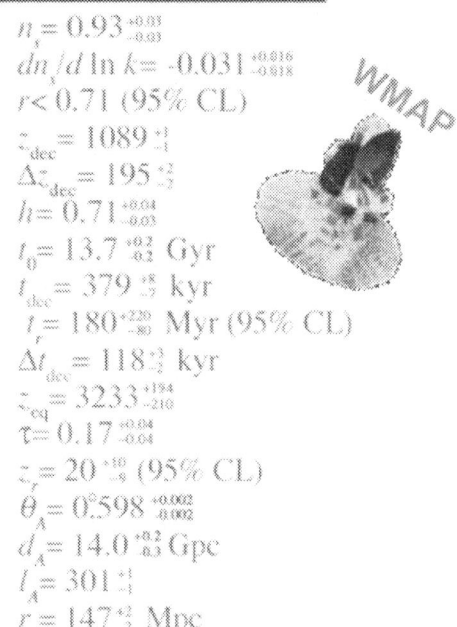

$\Omega_{tot} = 1.02^{+0.02}_{-0.02}$

$w < -0.78$ (95% CL)

$\Omega_\Lambda = 0.73^{+0.04}_{-0.04}$

$\Omega_b h^2 = 0.0224^{+0.0009}_{-0.0009}$

$\Omega_b = 0.044^{+0.004}_{-0.004}$

$n_b = 2.5 \times 10^{-7}{}^{+0.1\times10^{-7}}_{-0.3\times10^{-7}}$ cm^{-3}

$\Omega_m h^2 = 0.135^{+0.008}_{-0.009}$

$\Omega_m = 0.27^{+0.04}_{-0.04}$

$\Omega_\nu h^2 < 0.0076$ (95% CL)

$m_\nu < 0.23$ eV (95% CL)

$T_{cmb} = 2.725^{+0.002}_{-0.002}$ K

$n_\gamma = 410.4^{+0.9}_{-0.9}$ cm^{-3}

$\eta = 6.1 \times 10^{-10}{}^{+0.3\times10^{-10}}_{-0.2\times10^{-10}}$

$\Omega_b \Omega_m^{-1} = 0.17^{+0.01}_{-0.01}$

$\sigma_8 = 0.84^{+0.04}_{-0.04}$ Mpc

$\sigma_8 \Omega_m^{0.5} = 0.44^{+0.04}_{-0.05}$

$A = 0.833^{+0.086}_{-0.083}$

$n_s = 0.93^{+0.03}_{-0.03}$

$dn_s/d \ln k = -0.031^{+0.016}_{-0.018}$

$r < 0.71$ (95% CL)

$z_{dec} = 1089^{+1}_{-1}$

$\Delta z_{dec} = 195^{+2}_{-2}$

$h = 0.71^{+0.04}_{-0.03}$

$t_0 = 13.7^{+0.2}_{-0.2}$ Gyr

$t_{dec} = 379^{+8}_{-7}$ kyr

$t_r = 180^{+220}_{-80}$ Myr (95% CL)

$\Delta t_{dec} = 118^{+3}_{-2}$ kyr

$z_{eq} = 3233^{+194}_{-210}$

$\tau = 0.17^{+0.04}_{-0.04}$

$z_r = 20^{+10}_{-9}$ (95% CL)

$\theta_A = 0.598^{+0.002}_{-0.002}$

$d_A = 14.0^{+0.2}_{-0.3}$ Gpc

$l_A = 301^{+1}_{-1}$

$r_s = 147^{+2}_{-2}$ Mpc

Extra Space Dimensions?

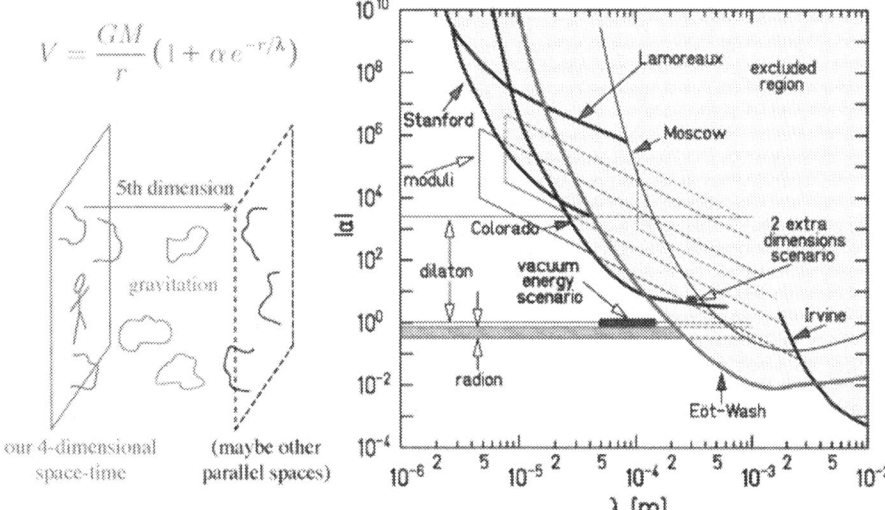

http://www-conf.slac.stanford.edu/ssi/2005/lec_notes/Esposito-Farese/esposito_Page_16_jpg.htm
[C.D. Hoyle *et al.*, Phys. Rev. **D70** (2004) 042004, hep-ph/0405262]

New limits of broader range than the one above were reported by Professor Esposito-Farese at SLAC SSI 2005 by Dr. Joshua Long.

New Limit and Projections

http://www-conf.slac.stanford.edu/ssi/2005/lec_notes/Long/long_Page_36_jpg.htm

On Aug 2, 2005, at 2:04 PM, I wrote from Panofsky Auditorium:

T. Berry just reported here at SLAC SSI 2005 no conclusive data as yet on extra space dimensions, but she is optimistic. Note my little formula, (not given by Berry BTW)

$$\Lambda_{zpf} = \left(\frac{\lambda}{L_p}\right)^n \left(\frac{L_p^2}{\lambda_{Compton}^4}\right) \tag{1.37}$$

for the random micro-quantum zero point vacuum fluctuation noise field that is the dark energy core of elementary particles as Bohm hidden variable micro-geons if there are n extra space dimensions of size λ, is derived as follows

$$\left(\frac{M_p}{M^*_p}\right)^2 = \left(\frac{\lambda}{L_p}\right)^n \tag{1.38}$$

This comes from

Compact Extra Dimensions

Solution to Hierarchy Problem: Weakness of gravity due to n "large" extra dimensions in which only gravitons propagate. Size R of extra dimensions related to unification scale M^* by:

$$R^n = M_p^2 / M^{*(2+n)} \qquad M^* \sim 1 \text{ TeV} \Leftrightarrow R \sim 1 \text{ mm for } n = 2$$

N. Arkani-Hamed, S. Dimopoulos, G. Dvali, Phys. Lett. B 429 (1998) 263

http://www-conf.slac.stanford.edu/ssi/2005/lec_notes/Long/long_Page_04_jpg.htm

$$\frac{G * M *_p^2}{\hbar c} \sim 1 \tag{1.39}$$

by definition. Therefore,

$$\alpha \equiv \frac{G *}{G} = \frac{M_p^2}{M *_p^2} = \left(\frac{\lambda}{L_p}\right)^n \tag{1.40}$$

Strong Short-Range Gravity

My Bohmian hidden-variable theory of the micro-geon requires (also proposed by Alexander Ya Burinskii in Moscow) $\alpha \sim 10^{40}$ & $\lambda \sim 10^{-13} cm$. Therefore $n = 2$. This set of parameters is way off the scale of the above data limit curves for direct experiment with torsion balances et-al. My idea here gives the universal Regge trajectories of the hadronic resonances of slope $\sim$ quantum of area $L *_p^2$ measured as $\alpha' \sim \left(1 Gev\right)^{-2}$ as rotating Kerr micro-geon black holes. The finite lifetime would come from Hawking radiation, or perhaps George Chapline's tiny "dark star" alternative model. The universality of the Regge slope is obviously the equivalence principle on the micro-scale. That's what I thought in 1973. I first published this kind of rough idea in 1973 in Herbert Frohlich's "Collective Phenomena", which prompted Abdus Salam to invite me to ICTP in Trieste, Italy then. Today we know that the large black hole is a large tangled string. That is

$$J \sim \alpha' E^2 \tag{1.41}$$

Space-time physics is local only because "classical" is an illusion that needs to be replaced by "macro-quantum" or "ODLRO" or "VEV" (Vacuum Expectation Value) from "spontaneous broken symmetry" or more generally "More is different" theory of bottom $\rightarrow$ up emergent order.

Obviously both

Bottom -> Up + Top -> Down

&

IT FROM BIT + BIT FROM IT

supplementing John Archibald Wheeler's

"Law without law"

Space-time physics is local because the macro-quantum Higgs VEV is local. This is a well-understood feature of Oliver Penrose's "ODLRO" of reduced micro-quantum density matrices (correlation functions).

The arrow of time, i.e. early universe post-Big Bang has low entropy that increases as our local universe's space expands in the multiverse of parallel other worlds, is because the formation of the Higgs VEV in the inflation from the false vacuum lowers the entropy as is well known in superfluid physics. On the other hand, we saw that Richard Gott et-al have a different explanation for the arrow of time in terms of a self-consistent loop in time at the very beginning of the self-creating megaverse of $\sim 10^{500}$ pocket universes populating every valley of the cosmic landscape. This also explains why our pocket universe is soon exceedingly fine-tuned to at least one part in about 10^{120}.

Professor "Rocky" Kolb from Fermi National Laboratory in Batavia, Illinois has kindly given me permission to use his slides from his SLAC SSI 2005 Lectures at Stanford University.

Is God Necessary?

Big-Bang

Robertson-Walker metric

$a(t) = $ cosmic scale factor

$k = 0, \pm 1$

$$ds^2 = dt^2 - a^2(t)\left(\frac{dr^2}{1-kr^2} + r^2 d\Omega^2\right)$$

$\Leftarrow G_{\mu\nu} = 8\pi G T_{\mu\nu} \Rightarrow$

Perfect-fluid stress tensor

$\rho = $ energy density

$p = $ pressure

$T^\mu_{\ \nu} = \text{diag}(\rho, p, p, p)$

http://www-conf.slac.stanford.edu/ssi/2005/lec_notes/Kolb1/kolb1_Page_04_jpg.htm

Field equations

$$\left(\frac{\dot{a}}{a}\right)^2 + \frac{k}{a^2} = \frac{8\pi G}{3}\rho \qquad H \equiv \frac{\dot{a}}{a} = \text{expansion rate}$$

$$\frac{\ddot{a}}{a} = -\frac{4\pi G}{3}(\rho + 3p) \qquad q \equiv -\frac{\ddot{a}}{a}\frac{1}{H^2} = \text{deceleration parameter}$$

$$T^{\mu\nu}_{\ \ ;\nu} = 0 \qquad \rho \propto a^{-3(1+w)} \qquad w = p/\rho$$

$T^{\mu\nu}$: fluids with different w

matter: $\qquad p_M = 0 \qquad w = 0 \qquad \rho_M \propto a^{-3}$

radiation: $\quad p_R = \rho_R/3 \qquad w = 1/3 \qquad \rho_R \propto a^{-4}$

vacuum: $\quad p_\Lambda = -\rho_\Lambda \qquad w = -1 \qquad \rho_\Lambda \propto a^0$

http://www-conf.slac.stanford.edu/ssi/2005/lec_notes/Kolb1/kolb1_Page_05_jpg.htm

Dynamics → Evolution

$$\left(\frac{\dot{a}}{a}\right)^2 + \frac{k}{a^2} = \frac{8\pi G}{3}\rho \qquad \rho = \rho_M(a) + \rho_R(a) + \rho_\Lambda(a) + \ldots$$

- **$a(t)$, $H(t)$, depend on matter/energy content**

- **$a(t)$ measurable via redshift $z \equiv a(0)/a(t)$**

- **Redshift z is a proxy for time or scale factor**

http://www-conf.slac.stanford.edu/ssi/2005/lec_notes/Kolb1/kolb1_Page_07_jpg.htm

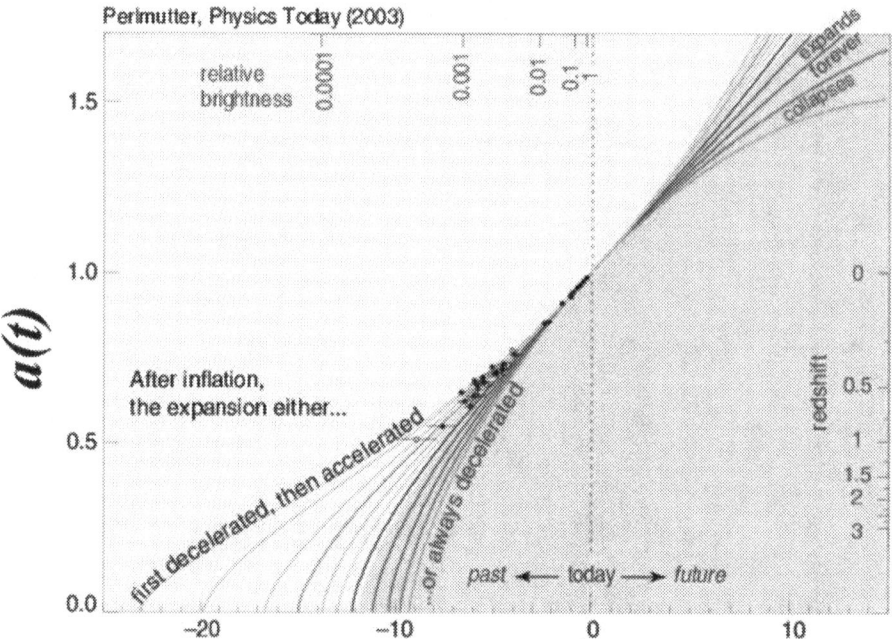

Time (billions of years from today)

The evidence for dark energy and dark matter comes from several independent observational technologies given by Rocky Kolb in the next slide.

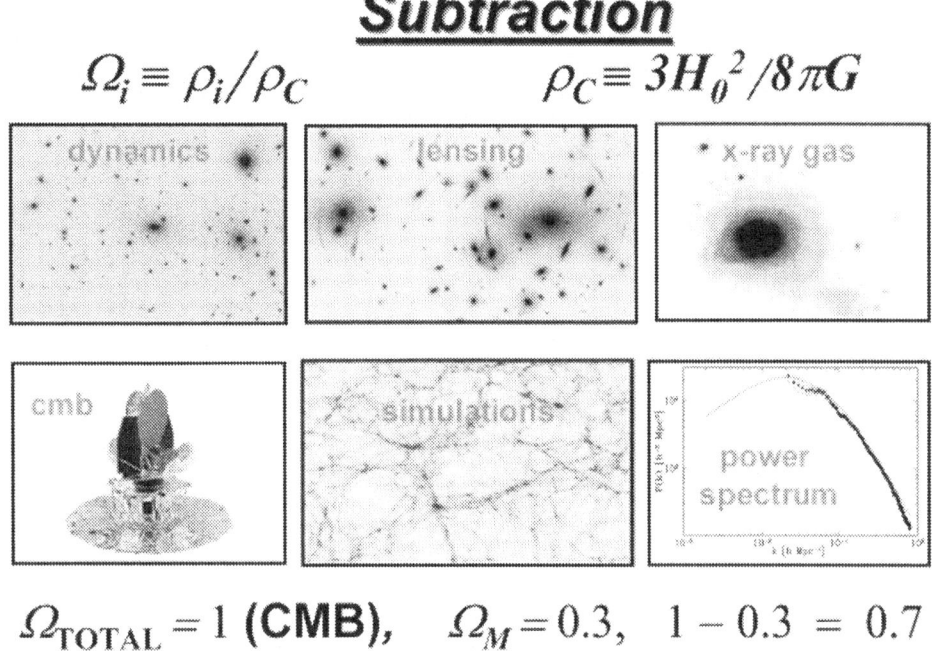

Subtraction

$$\Omega_i \equiv \rho_i/\rho_C \qquad \rho_C \equiv 3H_0^2/8\pi G$$

$$\Omega_{TOTAL} = 1 \text{ (CMB)}, \quad \Omega_M = 0.3, \quad 1 - 0.3 = 0.7$$

http://www-conf.slac.stanford.edu/ssi/2005/lec_notes/Kolb1/kolb1_Page_15_jpg.htm

What is most impressive is the convergence of all of these independent kinds of observations to a common meeting point shown in the next slide. Two is a random coincidence, but three or more is a cosmic conspiracy!

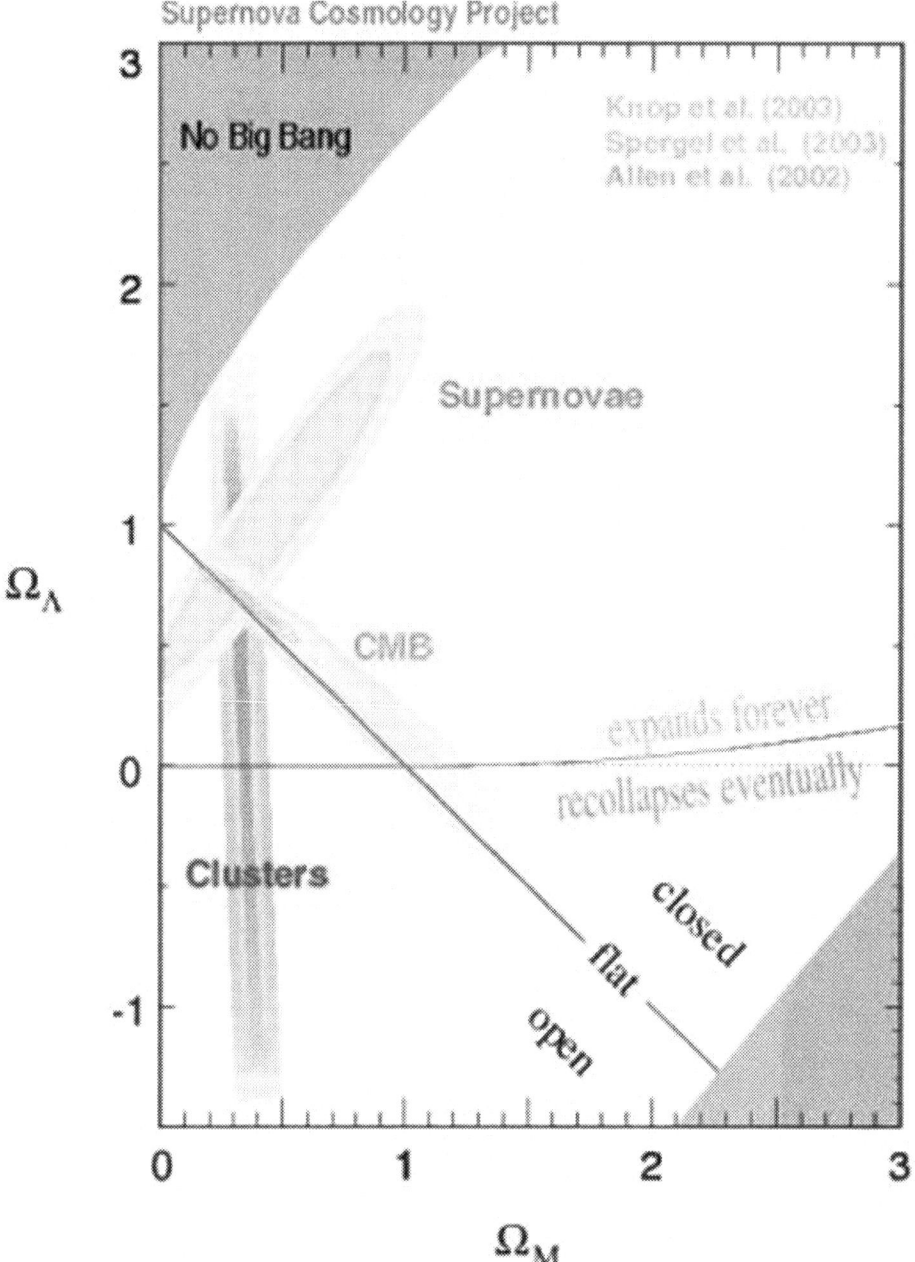

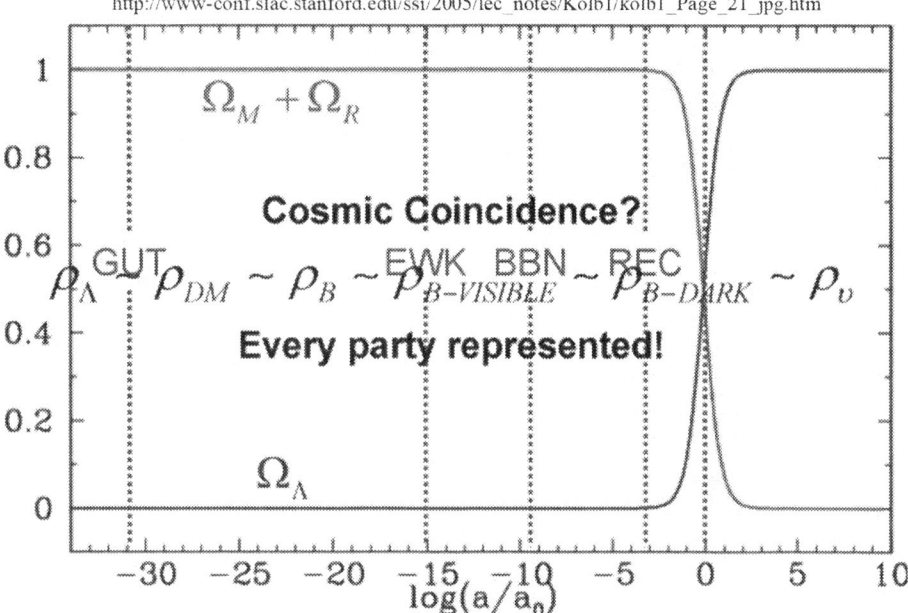

Cosmic coincidence

http://www-conf.slac.stanford.edu/ssi/2005/lec_notes/Kolb1/kolb1_Page_21_jpg.htm

The extreme right vertical dotted line needs to be moved over a little bit to the right in the above slide to show us where we are today with the rising dark energy curve intersecting the dotted line at approximately 0.7 while intersecting the falling dark matter curve at approximately 0.3. As our local universe in the multi-verse continues to expand and speed up, the vertical dotted line indicating "now" moves to the right, i.e. "0" for the logarithm, therefore

$$\frac{a(t)}{a_0} \equiv 10^{\log(a(t)/a_0)} \rightarrow 10^0 = 1 \tag{1.42}$$

OK, why are we today so close to the intersection point of the two curves for dark energy and dark matter? Also, are we stuck in this rotten development? Can we metric engineer our universe not to accelerate into isolated icy cosmic death? Are our children destined to shiver in the cold before the Twilight of the Gods (Gotterdammerung)?

http://www-conf.slac.stanford.edu/ssi/2005/lec_notes/Kolb2/Kolb2_Page_46_jpg.htm

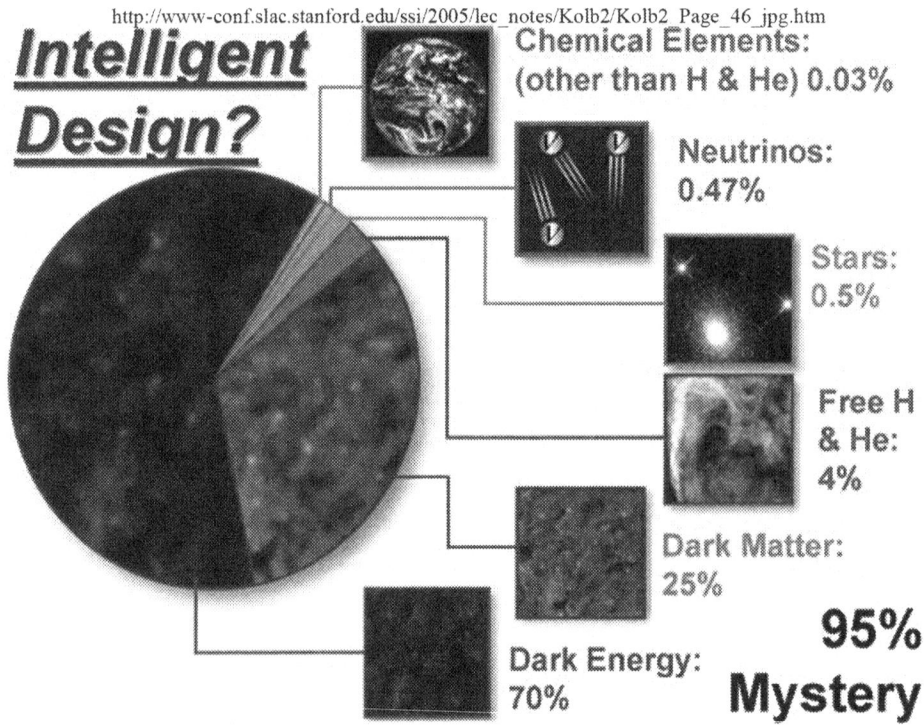

Is it a meaningless Tale of Sound and Fury told by The Idiot signifying nothing in the Weak Anthropic Principle (WAP) Darwinian selection Multiverse of Chaotic Inflation shown in A. Linde's

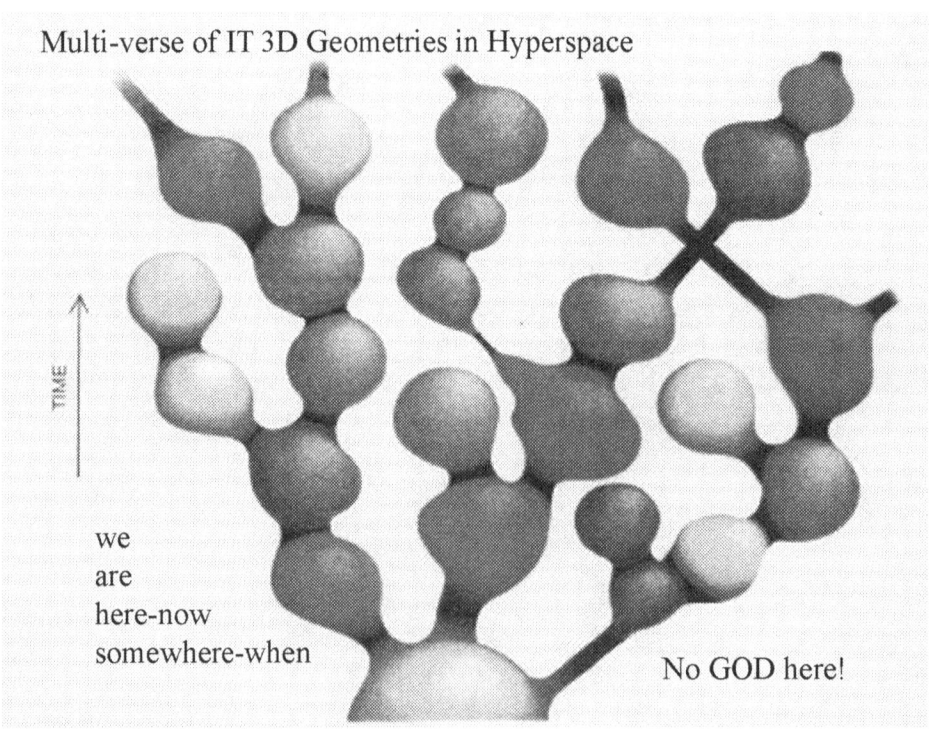

Here's what Lenny Susskind thinks:

"To Victor's question, 'Was it not God's infinite kindness and love that permitted our existence?' I would have to answer with Laplace's reply to Napoleon: 'I have no need of this hypothesis.' The Cosmic Landscape is my answer…" p. 15

Excerpts from John Walker's book review:

Sunday, March 12, 2006

Reading List: *The Cosmic Landscape*[cii]

*"Leonard Susskind (and, independently, Yoichiro Nambu) co-discovered the original hadronic string theory in 1969. He has been a prominent contributor to a wide variety of topics in theoretical physics over his long career, and is a talented explainer of abstract theoretical concepts to the general reader. This book communicates both the physics and cosmology of the "string landscape" (a term he coined in 2003) revolution, which has swiftly become the consensus among string theorists, as well as the intellectual excitement of those exploring this new frontier. The book is subtitled "String Theory and the Illusion of Intelligent Design" which may be better marketing copy—controversy sells—than descriptive of the contents. There is very little explicit discussion of intelligent design in the book at all except in the first and last pages, and what is meant by "intelligent design" is not what the reader might expect: design arguments in the origin and evolution of life, but rather the apparent fine-tuning of the physical constants of our universe, the cosmological constant in particular, without which life as we know it (and, in many cases, not just life but even atoms, stars, and galaxies) could not exist. Susskind is eloquent in describing why the discovery that the cosmological constant, which virtually every theoretical physicist would have bet had to be precisely zero, is (apparently) a small tiny positive number, seemingly fine tuned to one hundred and twenty decimal places "hit us like the proverbial ton of bricks" (p. 185)—here was a number which, not only did theory suggest should be 120 orders of magnitude greater, but which, had it been slightly larger than its minuscule value, would have precluded structure formation (and hence life) in the universe. One can imagine some as-yet-undiscovered mathematical explanation why a value is precisely zero (and, indeed, physicists did: it's called supersymmetry, and searching for evidence of it is one of the reasons they're spending billions of taxpayer funds to build the **Large Hadron Collider**), but when you come across a dial set with the almost ridiculous precision of 120 decimal places and it's a requirement for our own existence, thoughts of a benevolent Creator tend to creep into the mind of even the most doctrinaire scientific secularist. This is how the appearance of "intelligent design" (as the author defines it) threatens to get into the act, and the book is an exposition of the argument string theorists and cosmologists have developed to contend that such apparent design is entirely an illusion. The very title of the book, then invites us to contrast two theories of the origin of the universe: "intelligent design" and the "string landscape". So, let's accept that challenge and plunge right in, shall we? First of all, permit me to observe that despite frequent claims to the contrary, including some in this book; intelligent design need not presuppose a supernatural being operating outside the laws of science and/or inaccessible to discovery through scientific*

*investigation. The origin of life on Earth due to deliberate seeding with engineered organisms by intelligent extraterrestrials is a theory of intelligent design which has no supernatural component, evidence of which may be discovered by science in the future, and which is sufficiently plausible to have persuaded Francis Crick, co-discoverer of the structure of DNA, was the **most likely explanation**. If you observe a watch, you're entitled to infer the existence of a watchmaker, but there's no reason to believe he's a magician, just a craftsman. If we're to compare these theories, let us begin by stating them both succinctly:*

*   **Theory 1: Intelligent Design.**   *An intelligent being created the universe and chose the initial conditions and physical laws so as to permit the existence of beings like ourselves.*

*   **Theory 2: String Landscape.**   *The laws of physics and initial conditions of the universe are chosen at random from among 10^{500} possibilities, only a vanishingly small fraction of which (probably no more than one in 10^{120}) can support life. The universe we observe, which is infinite in extent and may contain regions where the laws of physics differ, is one of an infinite number of causally disconnected "pocket universes" which spontaneously form from quantum fluctuations in the vacuum of parent universes, a process which has been occurring for an infinite time in the past and will continue in the future, time without end. Each of these pocket universes, which, together, make up the "megaverse", has its own randomly selected laws of physics, and hence the overwhelming majority are sterile. We find ourselves in one of the tiny fraction of hospitable universes because if we weren't in such an exceptionally rare universe, we wouldn't exist to make the observation. Since there are an infinite number of universes, however, every possibility not only occurs, but occurs an infinite number of times, so not only are there an infinite number of inhabited universes, there are an infinite number identical to ours, including an infinity of identical copies of yourself wondering if this paragraph will ever end. Not only does the megaverse spawn an infinity of universes, each universe itself splits into two copies of itself every time a quantum measurement occurs. Our own universe will eventually spawn a bubble, which will destroy all life within it, probably not for a long, long time, but you never know. Evidence for all of the other universes is hidden behind a cosmic horizon and may remain forever inaccessible to observation.*

*   *Paging Friar Ockham! If unnecessarily multiplying hypotheses are stubble indicating a fuzzy theory, it's pretty clear which of these is in need of the razor!"*

The complete article can be found at:

> http://www.fourmilab.ch/fourmilog/archives/2006-03/000664.html

Well what do you think?

Index

Cap Weinberger, 102, 144
Capelas de Oliveira, 184
Capra, 29
Captain Ahab, 136
Captain Terauchi, 122
cargo cult, iii, 10
cargo cult pseudo-physics, 10
Carl Collins, 148
Carlo Suares, 56, 101, 107, 163
Carlos Barcelo, 180
Cartan, 3, 15, 55, 177, 181, 184, 192, 194, 197
Cartan 2-form, 181
Cartan forms, 181
Cartan frames, 55
Carter, 32
Casimir, 25, 43, 44, 46, 92, 140
Casimir effect, 92
Casimir force, 44
Casimir plates, 35
Catch 22, xi
causal loops, 21
causal retarded boundary, 152
causal structure, 101
c-bit, 165
CDM, 52, 203
CDMSII, 60
censorship, 106
center of mass, 151
center of mass quantum wave packet, 151
center of the Earth, 44
center-of-mass quantum state, 184
centrifugal acceleration, 206
Chadwell, 67
Chalmers, 85
chaotic eternal inflation, 177
chaotic inflation, 16, 218
Chapter 9, 4, 130, 131, 153
charge, 14, 15, 17, 35, 66, 86, 106, 132, 137,
    139, 140, 150, 169, 176, 178, 189
charge without charge, 178
Charge without charge, 168
charges, 13, 56, 134, 138, 169, 182, 188, 197,
    198
Charismatic Chain of Hermetic Adepts, 163
Chief of Station, 111
China, 154
Chinese, 90, 106
Chip, 111
Chipman's role, 111
chiral QCD, 188
Chris Carter, 108
Christ, 84
Christiansen, 178
Christopher "Kit" Green, 23
Christopher Green, 116
Christopher Marlowe, 5

Chuck Yeager, 168
Churchland, 83
CIA, 32, 66, 67, 86, 111, 112, 116, 130, 143,
    149, 154, 156
CIA Chief of Station, 111
CIA experiments, 23, 94
CIA/DIA Star Gate, 80
CIPA, 162
civil war, 10
classical brain, 89
classical curved space-time of gravity, 100
classical determinism, 90
classical electron, 139
classical field theory, 191
classical geometrodynamics, 247
classified projects, 154
Claude Shannon, 81, 88
clone a quantum state, 84
close encounter, 159
Close Encounters, 29, 110, 157
closed causal chains, 21
closed inexact, 181
closed inexact 1-form, 192
closed inexact 2-form, 192
closed loop, 192, 199
closed loops, 151, 169, 198
clusters of galaxies, 148
c-number fields, 194
Coast to Coast, 128
cockpit windows, 125
code, 32, 90
cohere, 86
coherence length, 169, 192
coherent order parameter, 82
coherent superfluid order, 132
coherent superposition, 62
coherent vacuum field, 192
cohomology, 185
coincidence problem, 202
Coit Tower, 102
cold dark matter, 150
Cold Dark Matter, 52, 203
cold fusion, 149, 162
cold metallic voice, 23
Cold War, 75, 110
cold war disinformation, 155
Colin Bennett, 28, 68
collapse, 83, 86, 166
collapse of the quantum wave function, 22
collapse of the state, 82, 96
collective modes, 59, 139
collective phenomena, 100
colliding branes, 16
Colm Kelleher, 116, 160, 247
Columbia University Psychology Department,
    110

224

228

impedance matching, 71, 127
implicate, 84, 85, 88
impossible maneuvers, 119
impossible maneuvers of the flying saucers, 151
impotent and obsolete, 67, 159
impulse propulsion, 120
inconsistent loops, 21
increasing thermodynamic entropy, 152
Independence Day, 115
independence from oil, 54
India, 154
India Daily, 74
individual events, 86, 94
inertial g-force, 53
infinite curvature singularity, 247
infinite energy in quantum electrodynamics, 150
infinite renormalization, 150
infinity of parallel universes next door, 152
inflation, 16, 17, 22, 55, 71, 155, 174, 175, 184, 185, 186, 188, 207, 212
inflation /arrow of time problem, 55
inflation bubble, 14, 169, 188
inflation field, 155, 177, 181, 188
Inflation Field, 151
inflation vacuum condensate field, 181
inflation vacuum phase transition, 186
inflationary critical phase transition, 178
inflationary vacuum phase transition, 12
information, 83, 84, 85, 86, 87, 88, 89, 90, 91, 92, 97, 105, 124, 165, 166, 167
information field controls quantum particles, 150
information loss down a black hole, 203
Ingo Swann, 23, 111, 130
ingroup, 166
initial cosmic black body density fluctuations, 45
inner consciousness, 94, 101, 165
inner core, 60, 104
Inquisition, 93
instantaneous signals, 106
Institute for Contemporary Studies, 144
Institute of Advanced Studies, 3
Institute of Theoretical Astronomy, 153
integer spin bosons, 58
Intelligence, 66, 67, 110, 111, 112, 126, 130, 131, 156, 164
intelligence community, 154
intelligent design, 45, 219
Intelligent Design, 17, 163
intelligent designer, 19
Intelligent Designer, iii, 57
intelligent extraterrestrials, 220
Intelligent Universe, 101
interferometer, 198
interferometers, 58
internal symmetry, 44
internal symmetry group, 44

International Centre for Theoretical Physics, 165
International Space Sciences Organization, 149, 155
Internet, 32
intolerance, 106
invariant, 10
inventors, 102
inverse square law of gravity, 120
Invisible College, 3
ionized plasma, 151
IR/UV, 100, 160
Ira Einhorn, 110
Iran, 154
Iran's nuclear weapons, 131
Iranian Air Force, 63
irreducible randomness, 165
Irreducible randomness, 94
irreversible, 105, 152, 166
Isaac Newton, 163
ISEP, 150
ISEP Corporation, 191
Isidor Isaac Rabi, 164
Islam, 162
isomer bomb, 143
Isomer weapons, 143
isomeric gamma ray weapons, 148
isotropy of the universe, 44
ISR, 120
Israel, 40, 131, 154
ISSO, 149, 162
ISSO Torsion Workshop, 162
IT, 82, 100, 105
IT FROM BIT, 179, 197
IT/BIT dualism, 105
Italian Physical Society, 12

J.A. Wheeler, 12
J.P. Vigier, 162
Jack Sarfatti, iii, 28, 35, 42, 63, 71, 126, 128, 138, 154
Jack Sarfatti and and Hal Puthoff, 28
Jacob Atabet, 110
Jacques Bergier, 3
Jacques Vallee, 3, 101, 107, 110, 116, 130, 151, 157
Jaffe, 131
Jagdish Mann, 102
Jahn-Teller effect, 146
James Jesus Angleton, 128, 160
Jane's Defence Weekly, 60, 149, 154
Jean Pierre Vigier, 165
Jean Reisser Nadal, 110
Jeffrey Mishlove, 85
Jeremy Bernstein, 102
Jesus Christ, 7
Jew, 6

238

End Notes

[i] http://www.physics.brocku.ca/etc/cargo_cult_science.html

[ii] John Archibald Wheeler at APS, April 2003, Philadelphia at age over 90.

[iii] See the book by Colm Kelleher and George Knapp discussed in this book that suggests we are under attack by a malevolent alien intelligence. This might explain the threats by Mahmoud Ahmadinejad,the President of Iran that could destroy most of the life on Earth in 2006. Is he a puppet of the Skinwalkers? See Howard Bloom's book, "The Lucifer Principle" to follow up on this line of inquiry.

[iv] MASINT is a science-intensive discipline that needs people/scientists well versed in the broad range of physical and electrical sciences. Such scientists cannot typically be professionally developed with the IC. They must come from academia fresh with scientific knowledge from experimentation and research. Nor can they continue to be "proficient" in their areas of expertise if they remain in government employ for an entire career. http://www.access.gpo.gov/congress/house/intel/ic21/ic21007.html

[v] "I think Einstein's idea that the presupposition of absolute time was at the core of the tremendous confusion and turmoil at the time is the core idea of special relativity, and his solution to insist on an experimental procedure to determine space-time differences between two point was the logical method to challenge that presupposition; the specific method he proposed (using light signals) is in my view canonical. Or do you have an alternative suggestion for how to do it? In this sense it is misleading to call it a convention. I think Einstein hit the nail on the head, and his fundamental method used then (insistence on operational definitions of basic concepts in physics) is one of the greatest revolutions in physics, far beyond just the SR or even GR. A lot of the current bullshit (some of it brilliant bullshit, admittedly) going on in physics (including the bombastic claims of string theorists of having found the theory of all even though they haven't made a single observable prediction) could have been avoided if physicists had taken this basic point more to heart and insisted on relating everything back to experience. Back to the SR situation specifically: with the 1905 paper, the confusion surrounding Michelson-Morley and the Lorentz transformation dissolved; and simplicity, beauty, and the accuracy of experimental predictions reigned again. I don't think Lorentz himself would have taken the position you are taking (once he had digested Einstein's approach), even though his name is associated with the position you are taking now because it was Lorentz's thinking BEFORE 1905. I haven't heard back from you regarding a defensible formulation of the supposed contradiction within SR itself that you have been arguing. I take it that we are then leaving it at that, and Einstein vs. Zielinski is settled. Phew, I am relieved ☺ Ph.D. UC physics-degreed George Weissmann to Paul Zielinski on Feb. 26, 2005

[vi] http://classiclit.about.com/library/bl-etexts/cmarlowe/bl-cmarlowe-faust.htm
http://www.charles-gounod.com/vi/extraits/midi/faust2.mid

[vii] Thanks to the allegedly late Gail Whittaker for reminding me
http://www.phinnweb.com/links/literature/borges/aleph.html

[viii] http://www.techgnosis.com"Rapture, rapture!"
http://math.boisestate.edu/gas/yeomen/web_opera/yeomen_21.html

[ix] R.U. Sirius talk show #4 at http://rusiriusradio.com/

[x] http://heP.bu.edu/abstracts/creminelli-feb-04.html
http://www-ctP.mit.edu/cosmo/y0304/arkani_abs.html
http://cfa-www.harvard.edu/cfa/tad/events/s_abstracts/12.html

[xi] http://xxx.lanl.gov/abs/gr-qc/0012094

[xii] A recent paper in Physics B by Samir Mathur, et-al suggests that the giant string inside the black hole removes the infinite curvature singularity of the Penrose-Hawking classical geometrodynamics and violates John Wheeler's conjecture that "the black hole has no hair. " My theory also can do that in principle without strings vibrating in extra space dimensions because of the negative quantum pressure assumed to be absent in the Penrose-Hawking space-time singularity theorems.

[xiii] http://www.physorg.com/news10682.html

[xiv] Hagen Kleinert, Free University of Berlin, Department of Physics

[xv] from song by Gillespie/Coots

[xvi] http://roswellproof.com/files/ramey2.jpg

[xvii] Note the opposite happens electromechanically. For example, in the Casimir plate effect the longitudinal zero point pressure perpendicular to the plates is negative causing the plates to squeeze together from the absent long wave modes. There is also an anti-gravity repulsion but it is too small to be noticed in that experiment. Therefore, this is counter-intuitive.

[xviii] arXiv:quant-ph/0506141 v1

[xix] http://www.greylodge.org/occultreview/glor_005/drgreen.htm

[xx] http://www.scientainment.com/pchant.html

[xxi] From: Todd R. LePine, MD
To: Susan Waitt
Subject: RE: Question
Date: Sat, 9 Jul 2005 12:01:26 -0400
"Hi Susan,
I would be glad to write a few lines about the incident with Jack. From what I remember Jack was asking Steve Greer a couple of questions such as who were his sources for the scalar weapons and the alternative energy sources when he was confronted by Jim Berry who tried to grab the microphone from Jack, who resisted in giving up the microphone. In fact he asked the audience if they wanted him (Jack) to proceed in his asking the probing questions and there was a consensus to continue, but Jack was not allowed and things came to an abrupt closure. I was vociferously defending out loud Jack's right to pose questions to Steve Greer so as to glean what was truth and what was speculation in what he had to say during his talk."
Todd re: Incident at the La Fonda Hotel, Santa Fe, New Mexico, Sunday ~ 6pm April 24, 2005.

[xxii] On May 16, 2004, at 6:58 PM, Jack Sarfatti wrote: We had a meeting today with Ken Shoulders, his son Steve, Creon Levit and Faustin Bray. Ken told me a lot about his history with Hal Puthoff and Jupiter Technologies with Bill Church. They had Yakir Aharonov and David Finkelstein as consultants - other top guys as well. I went over Hal's Earthtech Site with Creon a little on the experiments.
http://www.earthtech.org/experiments/index.html Hal is doing good stuff there, basically checking out free energy claims same as we were doing at ISSO. Hal does a competent and honest job there and I support his work there and also in his old RV stuff with Targ. Creon is familiar with that work. My disagreement with Hal is strictly over only two theoretical issues his PV theory and his ZPE theory with Bernie Haisch as a theory for metric engineering. http://www.earthtech.org/publications/index.html

[xxiii] Colin Bennett, http://www.combat-diaries.co.uk/ "Politics of the Imagination" (the life, work, and ideas of Charles Fort) awarded Best Biography for 2002

[xxiv] http://www.disinfo.com/archive/pages/article/id773/pg1/

[xxv] Jack Sarfatti to Hal Puthoff on May 13, 2004: Send me your statement on "vacuum coherence" by tomorrow and I will include it. Otherwise, what I say is in the book in hard copy. I have given you ample time! I am saying in the book for the record:
1. You never mention "vacuum coherence" in any of your ZPE or PV papers.
Is this false?
2. You have never given a precise explanation either in English or in math of what you mean by "vacuum coherence" and how it differs from my precise use of the same term.
3. You have never shown how the QED concept of PV i.e. virtual e+e- pairs relates to either the HRP model or the PV model.
4. You give no way to overcome the G/c^4 barrier for practical metric engineering.
So here is your chance to correct the record. I have given you plenty of time and note:

[xxvi] http://www.nature.com/nsu/040503/040503-7.html

[xxvii] http://www.abc.net.au/science/news/stories/s1567144.htm

[xxviii] http://news.bbc.co.uk/2/hi/science/nature/4679220.stm

[xxix] The work done is independent of the path in physical space when there is no friction or time and velocity dependence of the force.

[xxx] Note internal symmetry of the strong force is SU(3) for the 3D harmonic oscillator.
http://www.iop.org/EJ/abstract/0305-4470/37/7/022

[xxxi] Allusion to Erwin Schrodinger's quip about "those infernal quantum jumps."

[xxxii] http://online.itp.ucsb.edu/online/smatter_m06/chan/oh/02.html

xxxiii http://www.nsa.gov/ufo

xxxiv http://brumac.8k.com/cia_explaination.html

xxxv http://www.nsa.gov/ufo/ufo00020.pdf

xxxvi as an electronic book at Filament Books (http://www.filamentbooks.com) and probably as a print-on-demand shortly thereafter from the UFO Magazine website (http://www.ufomag.com) and the Shadow Lawn Press website (http://www.shadowlawnpress.com). I will also try to offer it as an encrypted .pdf for download as well with both an audio and video supplement embedded in the text.

xxxvii http://brumac.8k.com/bio.html

xxxviii http://www.unicusmagazine.com/p2.htm

xxxix http://www.indiadaily.com/editorial/1944.asp

xl There is no absolute instant of time, i.e. no absolute simultaneity in globally flat special relativity without any gravity at all. You can slice 4D space-time into a 3D spacelike + 1D timelike split in an infinity of ways. Spacelike means outside the light cone and timelike means inside the light cones. All the light cones, one at each point in 4D space-time do not tilt relative to each other. When there is a relative tilt between neighboring light cones you have tidal relative acceleration curvature between two tiny point test particles each in free float motion along timelike geodesics. Special relativity only works locally in curved space-time with gravity. That's Einstein's "equivalence principle." Even different space-time points lose their individuality in general relativity. This leads to "relationalism."

xli Stuart Hameroff's term.

xlii The same as Carlo Suares's "dual cosmic flow" for the generation of inner consciousness.

xliii So did Sir Michael Berry at a different time. Creon Levit was there.

xliv http://qedcorP.com/book/psi/hitweapon.html

xlv http://www.astro.cf.ac.uk/groups/relativity/papers/abstracts/miguel94a.html

xlvi http://www-physics.lbl.gov/~stapp/presponse.txt

xlvii http://www.fourmilab.ch/rpkp/ http://www.fourmilab.ch/rpkp/staPp.html

xlviii http://xxx.lanl.gov/abs/quant-ph/0203049

xlix http://sped2work.tripod.com/satori.html

l http://www.bestjudo.com/brsatori3.shtml

li Convicted of the murder of Holly Maddox. The story is in my book "Space-Time and Beyond II." Einhorn was working with Vallee on extending DARPA net to become the Internet.

lii http://www.sbvpartners.com/vallee.html

liii http://www.nidsci.org/bios/vallee.html

liv http://www.edge.org/3rd_culture/susskind03/susskind_index.html

lv http://www.paranoiamagazine.com/nothing.html

lvi http://www.astro.cf.ac.uk/groups/relativity/papers/abstracts/miguel94a.html

lvii http://bcornet.homestead.com/files/weirdsound.htm

lviii www.earthtech.org/publications/Mitre%20Conference.pdf

lix http://www.ldolphin.org/hill.html

lx http://www.edge.org

lxi A test particle is passive. It is acted upon without back-action on what is acting on it. This is the approximation of action without reaction.

lxii http://www.space-drives.org/wwforwardrl.html
"Negative Mass in General Relativity", H. Bondi, Reviews of Modern Physics, Vol 29, July 1957, pp 423-428
"Negative and Imaginary Proper Masses", Y.P Terletskii, Paradoxes in the Theory of Relativity, Plenum, New York 1968, Chapter VI pp 83-115
"Negative Masses and the Energy-Sources of the Universe", Y.P. Terletskii, Experimentelle Technik der Physik, Vol 29 April 1981 pp 331-332

lxiii http://english.pravda.ru/science/19/94/377/12778_weapons.html

lxiv http://qedcorp.com/APS/ice9.wav

[lxv] Figure from Alcubierre's warp drive paper
http://www.astro.cf.ac.uk/groups/relativity/papers/abstracts/miguel94a.html
http://www.npl.washington.edu/AV/altvw99.html

[lxvi] E-mails from Doty make his participation in the actual writing of the book unclear. Doty seems to repudiate much of what Collins has in the book.

[lxvii] http://wlap.physics.lsa.umich.edu/umich/mctp/conf/2001/sto2001/okun/real/f012.htm

[lxviii] http://www.arxiv.org/abs/hep-ph/9910333

[lxix] http://english.pravda.ru/science/19/94/379/12737_weapons.html

[lxx] Rev. Mod. Phys. 59, S1–S201 (1987)

[lxxi] http://www.sfgate.com/cgi-bin/article.cgi?file=/chronicle/archive/1997/08/17/SC46892.DTL

[lxxii] http://www.firstgov.gov/

[lxxiii] http://www.acq.osd.mil/dpap/Docs/FY01RPT.doc

[lxxiv] http://english.pravda.ru/science/19/94/377/12778_weapons.html

[lxxv] http://www.newscientist.com/news/news.jsp?id=ns99994049

[lxxvi] http://www.nidsci.org/bios/bigelow.html

[lxxvii] http://www.tampabaycoalition.com/files/720ManBehindSavageNation.htm

[lxxviii] http://www.heP.upenn.edu/~max/
http://www.ldolphin.org/hill.html

[lxxix] David Langlois at GR 17 (2004)

[lxxx] http://arxiv.org/abs/gr-qc/0009013

[lxxxi] Term coined by Brian Greene in his popular books.

[lxxxii] Christiansen, W.A., Ng, Y. Jack, and van Dam, H. Probing Spacetime Foam with Extragalactic Sources. *Physical Review Letters*. 96, 051301 (2006).

[lxxxiii] arXiv:gr-qc/0106002 v1 1 Jun 2001

[lxxxiv] J. Bardeen, L.N. Cooper & J.R. Schrieffer, "Microscopic Theory of Superconductivity" Phys. Rev. 106, 162-164 (1957).

[lxxxv] http://www.ime.unicamp.br/rel_pesq/2005/rp56-5.html

[lxxxvi] N. Cimento B 26, 577-591, 1975

[lxxxvii] http://arxiv.org/abs/astro-ph/0302131

[lxxxviii] See Gary Bekkum's Star Stream site on the potential applications of Ken Shoulders "charge clusters" here modeled as mesoscopic EVOs "Exotic Vacuum Objects" where a strong attractive gravity induced by zero point vacuum energy positive pressure inside the shell of charge stabilizes it. This has nothing to do with the electromagnetic Casimir force that can also be included into the model, but will not change the results too drastically.

[lxxxix] F.Wilczek, "Fractional Statistics and Anyon Superconductivity." (World, 1990)

[xc] Sidney Coleman, "Aspects of Symmetry", 1973 Erice Lecture

[xci] T.W.B. Kibble, Lorentz Invariance and the Gravitational Field, J. Math. Phys. 2, 212-221 (1961), reprinted in "Gauge Theories in the Twentieth Century" J.C. Taylor (Imperial College Press, 2001)

[xcii] Hagen Kleinert, http://www.physik.fu-berlin.de/~kleinert/

[xciii] R. Penrose & W. Rindler, "Spinors and space-time," Cambridge (1984)

[xciv] G. Toulouse & M. Kleman, Principles of a Classification of Defects in Ordered Media, J. De Physique, 37, June 1976 pp. 149-151

[xcv] J. Baez & J.P. Muniain, "Gauge Fields, Knots and Gravity", pp 130-152 (World, 1994)

[xcvi] R. Becker, Thesis, U. Alabama, Huntsville, 1994

[xcvii] Leonard Susskind, "Cosmic Landscape" (2006)

[xcviii] http://www.physics.ucla.edu/hep/dm06/talks/gilmore.pdf

[xcix] There is an alternative zero point energy explanation of the same data in which the 9 km/sec inference from the virial theorem does not occur. Note that Gilmore reports 50 million solar masses equivalent for the

dark matter halo, that's $5 \times 10^7 \times 2 \times 10^{33} = 10^{41}$ grams. One parsec is 3×10^{18} cm . So that's an average mass density of $\sim 10^{41}$ grams$/(10^{21})^3$ cm$^3 \sim 10^{-22}$ grams per cc. The critical effective mass of our pocket universe that would make 3D space flat on the large scale is $\sim 10^{56}$ gram equivalent in a Hubble sphere of $\sim 10^{28}$ cm giving a density that is approximately $10^{56}/(10^{28})^3$ grams per cc $\sim 10^{-28}$ grams per cc. Therefore, the dark matter density on the small scale is about a million times greater in absolute value then is the cosmological dark energy density. How does this fit with George Chapline's scenario for dark matter in terms of a gas of small dark energy stars? http://xxx.arxiv.org/pdf/astro-ph/0503200

[c] http://www.physics.ucla.edu/hep/dm06/talks/primack.pdf

[ci] Dr. Eric Davis in STAIF 2006 has written on negative zero point energy in the Casimir effect. It is true that in the parallel plate configuration that the relative sizes of the ZPF stress-energy diagonal are -1, +1, +1, -3 so that the longitudinal pressure is negative. The quantum electromechanical suction causes the plates to attract, on the other hand this same negative pressure causes direct gravitational repulsion. However, the relative size of the two opposing simultaneous effects is (QED String Tension)/(Gravity String Tension) = (Plank Area) (Plate Area) / (Plate Separation) ^ 4 >> 1. Note, for isotropic distributions, that negative energy density gives gravity repulsion only if w > -1/3, otherwise it gives attraction. Similarly positive energy density gives gravity repulsion only for w < -1/3 otherwise it gives attraction. Lorentz invariance and the equivalence principle minimal gravity coupling imply w = -1 for isotropic distributions of zero point energy from all quantum fields.

[cii] Susskind, Leonard. *The Cosmic Landscape*. New York: Little, Brown, 2006. ISBN 0-316-15579-9.